ENERGY POLICY FOR PEACE

ENERGY POLICY FOR PEACE

Edited by

DANIEL M. KAMMEN
Energy and Resources Group, University of California, Berkeley, CA, United States

HISASHI YOSHIKAWA
Institute for Future Initiatives and Graduate School of Public Policy, The University of Tokyo, Tokyo, Japan

KENSUKE YAMAGUCHI
Graduate School of Public Policy, The University of Tokyo, Tokyo, Japan

Academic Press is an imprint of Elsevier
125 London Wall, London EC2Y 5AS, United Kingdom
525 B Street, Suite 1650, San Diego, CA 92101, United States
50 Hampshire Street, 5th Floor, Cambridge, MA 02139, United States
The Boulevard, Langford Lane, Kidlington, Oxford OX5 1GB, United Kingdom

ISBN: 978-0-12-817350-3

For Information on all Academic Press publications
visit our website at https://www.elsevier.com/books-and-journals

Publisher: Candice G. Janco
Acquisitions Editor: Kathryn Eryilmaz
Editorial Project Manager: Zsereena Rose Mampusti
Production Project Manager: Fahmida Sultana
Cover Designer: Christian J. Bilbow

Typeset by MPS Limited, Chennai, India

Contents

List of contributors

Venkatachalam Anbumozhi Economic Research Institute for ASEAN and East Asia, Jakarta, Indonesia

Daniel Del Barrio-Alvarez Department of Civil Engineering, The University of Tokyo, Tokyo, Japan

Jairo Alberto Garcia-Riveros Social Innovation Science Park, Corporación Universitaria Minuto de Dios - UNIMINUTO, Colombia

Daniel M. Kammen Energy and Resources Group, University of California, Berkeley, CA, United States; Goldman School of Public Policy, University of California, Berkeley, CA, United States

Noah Kittner Department of Environmental Sciences and Engineering, University of North Carolina at Chapel Hill, Chapel Hill, NC, United States; Department of City and Regional Planning, University of North Carolina, Chapel Hill, NC, United States

David Mozersky Energy Peace Partners, San Francisco Bay Area, United States

Masako Numata International Projects Division, The Nippon Foundation, Tokyo, Japan; MbSC2030, The University of Tokyo, Tokyo, Japan

Samira Siddique Energy and Resources Group, Renewable and Appropriate Energy Laboratory, University of California, Berkeley, CA, United States

Masahiro Sugiyama Institute for the Future Initiative, The University of Tokyo, Tokyo, Japan

Nguyen Quy Tam Fulbright School of Public Policy and Management, Fulbright University Vietnam, Vietnam

Jose Luis Wong-Villanueva Department of Urban Engineering, School of Engineering, The University of Tokyo, Tokyo, Japan

Kensuke Yamaguchi Graduate School of Public Policy, The University of Tokyo, Tokyo, Japan

Introduction

Chapter outline

1.1 Introduction

Energy use—which encompasses everything from the consumption of fuel for cooking to the use of renewable energy for electrification or heating—has long been associated with improvements in development and social outcomes. Aggregated analyses of national outcomes indicate that life expectancy and literacy rates exhibit a positive correlation with increases in commercial energy consumption per capita, up to a saturation point (Goldemberg, 1996; Jacobson, Milman, & Kammen, 2005). Strong positive correlations have additionally been observed between energy use per capita and economic development measures such as GDP per capita and the Human Development Index (Lee, Miguel, & Wolfram, 2017). These positive relationships with GDP per capita continue to persist when energy use is defined solely as electricity use, and the sample of countries under consideration is limited to developing nations (Stern, Burke, & Bruns, 2016). The energy development nexus is, of course, well known, and in functioning states, core to both theoretical and practical elements of development policy.

There is also, however, another side of the energy development nexus, in which the rise of distributed renewable energy can greatly facilitate: the role of sustainable energy in crisis settings, state building, and in empowering the most vulnerable people and communities in the conflict, refugee, and other critically marginalized settings. This book explores cases of the emerging field of crisis intervention through the unique properties of clean energy. To see how dramatically different clean energy for crisis intervention and addressing the needs in *failed* states can be, we need to both

Energy Policy for Peace. DOI: https://doi.org/10.1016/B978-0-12-817350-3.00011-0

set the context by looking more closely at traditional ideas of development and to motivate the chapters in this book.

For both traditional energy systems and in crisis and relief setting, the place to begin is by exploring the outcomes of energy services (Goldemberg, 1996; Lee et al., 2017). In each setting, electricity generation does not necessarily translate to access to reliable electricity, and evidence exists to suggest that blackouts in electrified areas with low reliability increase the vulnerability of women and girls to sexual violence (E4SV, 2015). Traditional institutional approaches to electrification may also serve to simply preserve inequities or even exacerbate them: expansion of grid access, which is typically associated with high capital costs, tends to favor wealthier communities. Such prioritization can further entrench existing inequities in power and economic status. Inequities may also be reinforced when gender dynamics, and their intersections with energy use, are not considered. In rural and peri-urban areas of many developing countries, for example, women and children tend to be the primary actors engaged with fuel collection and household uses of energy.

Collectively, the dynamics above suggest that initiatives to promote energy and electricity access must involve a broader spectrum of considerations than the simple physics of energy provision. More nuanced approaches to promoting development outcomes have long been advocated. Both classic and controversial works of sociology and development, from *Seeing Like a State* (Scott, 1998) to *Dead Aid* (Moyo, 2009), highlight that international aid and socioeconomic programs directed at improving development outcomes do not indubitably translate to reductions in poverty or increases in economic growth within the recipient countries. Such efforts are often well intentioned, but suffer from the lack of an informed approach. Within the sphere of technology adoption is the widely cited example of cookstoves. Noting that traditional methods of cooking using fuelwood or traditional charcoal-burning stoves produce large amounts of indoor pollution and contribute substantially to mortality, development actors have long sought to replace such options with cleaner-burning stoves. However, scholars such as Emma Crewe (1997) have documented how initial improvements in cookstove technologies were largely promulgated by expatriates and the technical expertise of engineers, without considering the needs and expertise of the local cooks who would actually be using the stove technologies. The take-up of more technically efficient stoves in Nepal, Gambia, Zimbabwe, India, and other countries in which the stoves were introduced was thus very low. For example, in Nepal, Fiji, and Guatemala, the newer, well-insulated, and fuel-efficient stoves were unpopular, because the use of stoves for space heating was more important than any gains in household welfare from fuel conservation (Crewe, 1997). Today, new actors are entering both the peacetime power sector space with mini-grid and other distributed products, and in relief and crisis settings, the "pop-up" possibility of remote power open opportunities for field clinics, refugee services, and other immediate needs can, *in theory*, be met in ways that were not previously possible in crisis settings.

1.2 Energy access in the development

Energy use has long been associated with improvements in development and social outcomes. For example, decreases in illiteracy and infant mortality rates have been observed to correlate with increases in energy use up to a saturation point (Goldemberg, 1996). As noted in the paper's introduction and shown in Fig. 1.1,

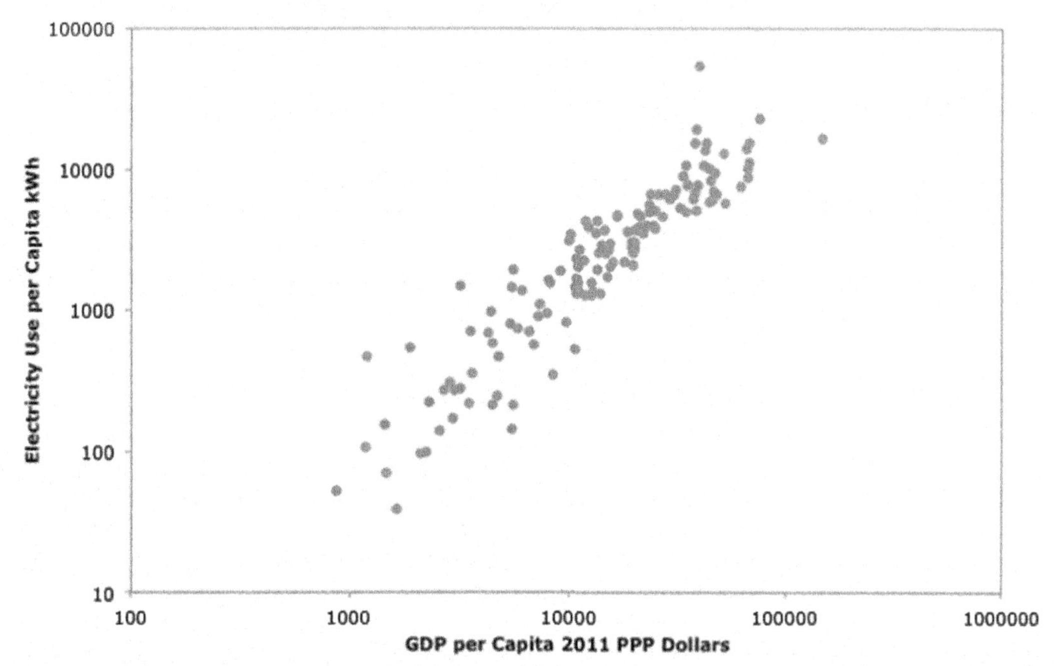

Figure 1.1 Per capita electricity use and per capita GDP (2014). From Stern, D. I., Burke, P. J., & Bruns, S.B. (2016). The Impact of Electricity on Economic Development: A Macroeconomic Perspective. EEG State-of-Knowledge Paper Series.

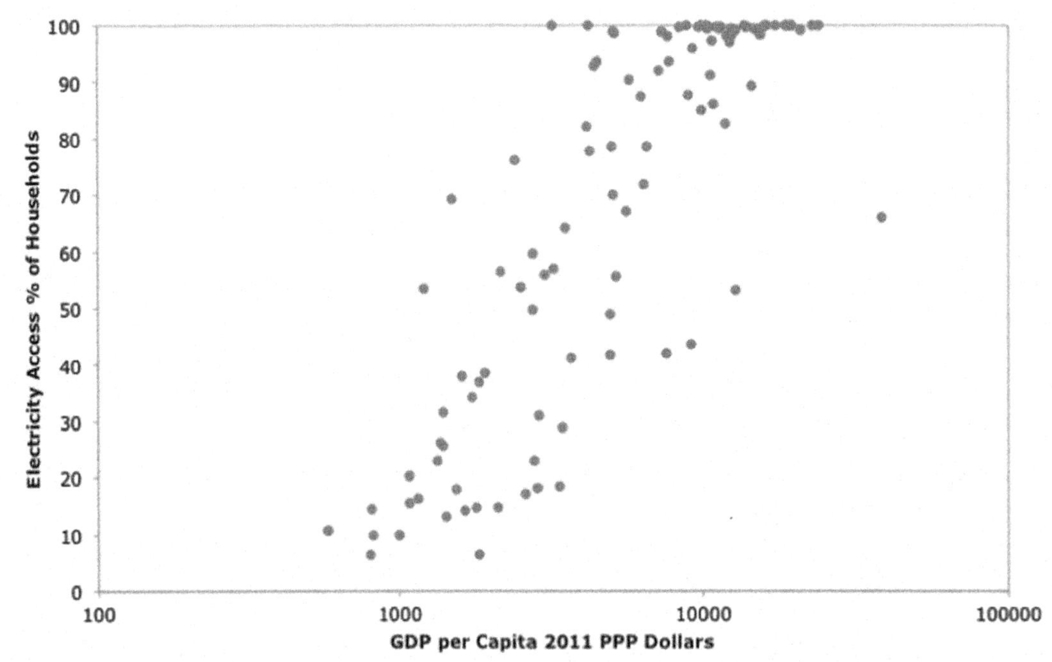

Figure 1.2 Electricity access (circa 2012) and GDP per capita (2014) for developing countries. From Stern, D. I., Burke, P. J., & Bruns, S.B. (2016). The Impact of Electricity on Economic Development: A Macroeconomic Perspective. EEG State-of-Knowledge Paper Series.

positive correlations are also observed between macroeconomic outcomes, such as per capita GDP, and the more specific energy use of electricity (Stern et al., 2016). Fig. 1.2 illustrates that a strong positive correlation also holds when the sample under consideration is limited to developing countries and looks at per capita GDP and the percent of households with electricity access (Stern et al., 2016).

A number of similar associative patterns can be observed at the nexus of energy and development, with fragile and conflict-affected situations. Morris (2017), for example, posits that "most fragile and conflict-affected situations have significantly worse development outcomes relating to energy." Analyzing electricity access in countries the UK's Department for International Development (DFID) identifies as priorities to receive funding; Morris highlights that "on average 43% of the populations in the 21 priority countries DFID considers as fragile and conflict-affected situations (FCAS) had access to electricity in 2014 (falling as low as 4% in South Sudan), whereas on average 58% of the populations in DFID's 7 non-FCAS priority countries had access to electricity" (Morris, 2017; World Bank, 2017).[1] Morris (2017) additionally cites research from the World Bank Group, Doing Business, which suggests that "whereas it takes on average 93 days to obtain a permanent electricity connection in DFID's 7 non-FCAS priority countries, this increases by over 47% to 137 days for DFID's FCAS countries."

The potential connections at the energy, development, and conflict nexus have been of increasing interest to actors in the aid community, as well as aid donors, including the UK's Department for International Development (DFID) and the World Bank. In its 2013 Energy Sector Directions Paper, for example, the World Bank posits that "providing electricity may be especially important in fragile and conflict-affected states, where resumption of electricity supply can be important in restoring confidence in the government, strengthening security and reviving the economy" (World Bank, 2013). Morris (2017) thus characterizes the energy sector as potentially "constitut [Ing] a central economic dimension of the so-called 'conflict trap': a sub-optimal equilibrium whereby poor performance in the energy sector therefore not only results from violence, but may also be one factor that creates the structural conditions for a continuation of violence."

Humanitarian aid and peacekeeping experts also recognize the role that energy access, particularly to renewable energy, may have for increasing resiliency in peace and recovery efforts. Mozersky, cofounder of Energy Peace Partners, highlights that "if you look at climate vulnerable areas, conflict risk maps and energy poverty, there's a very strong overlap on all three of these indexes" (Fleming, 2018). Fig. 1.3 provides one illustration of this overlap, by comparing the 40 most vulnerable countries separately ranked on the Fragile States Index (FSI) and the Notre Dame Global Adaptation Initiative Country Index (ND-GAIN), to countries where 60% or less of the population had electricity access, as of 2014, per the World Bank's Global Electrification Database. The FSI and ND-GAIN indices,

[1]DFID, the Department for International Development, is a department of the UK government with the responsibility of administering overseas aid. FCAS is an acronym referring to Fragile and Conflict-Affected States. Morris lists the 21 DFID FCAS priority countries as: Afghanistan, Bangladesh, Myanmar (Burma), Democratic Republic of Congo, Ethiopia, Kenya, Liberia, Malawi, Nepal, Nigeria, Occupied Palestinian Territories, Pakistan, Rwanda, Sierra Leone, Somalia, South Sudan, Sudan, Tajikistan, Uganda, Yemen, and Zimbabwe. However, the Occupied Palestinian Territories are excluded from the averages he calculates, due to a lack of data availability. Meanwhile, the seven DFID non-FCAS priority countries are: Ghana, India, Kyrgyzstan, Mozambique, South Africa, Tanzania, and Zambia.

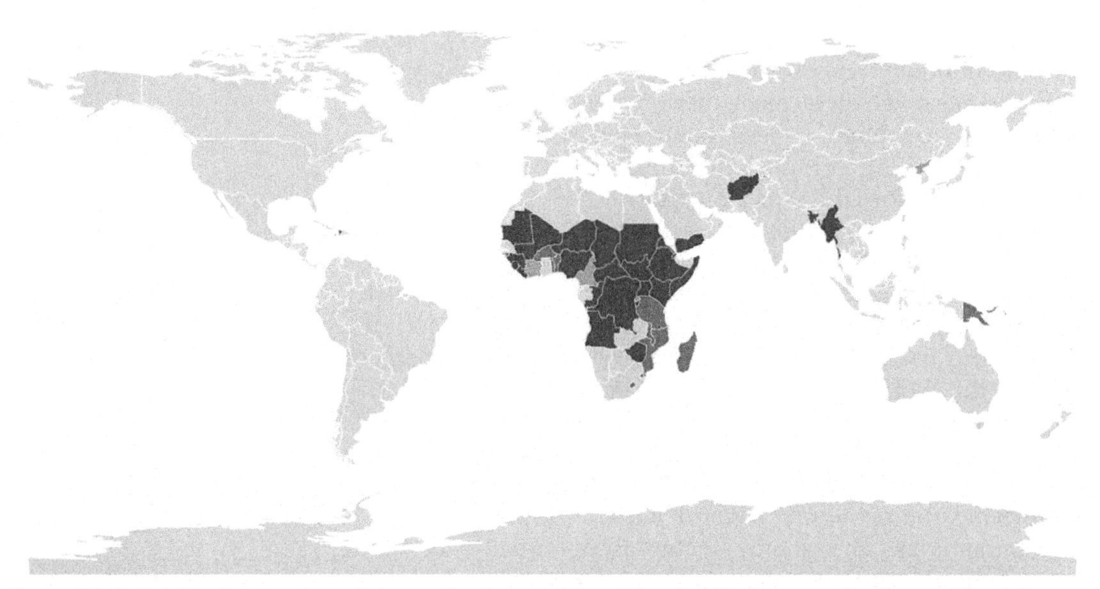

Figure 1.3 A high-level comparison of the overlap between countries classified as experiencing conflict risk, climate vulnerability, and energy poverty. Data are from the 2016 Fragile States Index, developed by the Fund for Peace, the 2015 ND-GAIN Country Index, from the University of Notre Dame, and the World Bank's Global Electrification Database from 2014. Thresholds used are the 40 most vulnerable countries listed on the FSI and ND-GAIN, and countries where 60% or less of the population has access to electricity ($n = 47$). Produced with David Mozersky & Daniel Kammen of Energy Peace Partners.

respectively, provide a measure of conflict risk and vulnerability to the effects of climate change.

However, effectively addressing the issue of energy access—particularly in developing and fragile and conflict-affected situations—requires a consideration of several intersecting dimensions. In a departure from earlier simplifications, energy access is no longer considered in the binary terms of whether someone does or does not have access. For example, in speaking about energy, types are important—energy access can refer to access to household fuels, including solid fuels such as wood and charcoal, or liquid fuels such as kerosene and liquefied petroleum gas (LPG); access to electricity; or to mechanical power. Even when narrowed to electricity more specifically, potential dimensions to consider include the structure or system of energy infrastructure (e.g., whether it is centralized or decentralized); the groups to whom access to provided (e.g., households, businesses/productive end uses, and/or to the community); the level of access provided; the quality and reliability of service; the fuel mix; costs for access; and available financing options.

Fig. 1.4, which depicts a Multi-tier Matrix for Access to Household Electricity Supply from the World Bank's Energy Sector Management Assistance Program (ESMAP), provides one example of how a more nuanced understanding of access has

Multi-tier Matrix for Measuring Access to Household Electricity Supply

ATTRIBUTES		TIER 0	TIER 1	TIER 2	TIER 3	TIER 4	TIER 5
1. Peak Capacity	Power capacity ratings[27] (in W or daily Wh)		Min 3 W	Min 50 W	Min 200 W	Min 800 W	Min 2 kW
			Min 12 Wh	Min 200 Wh	Min 1.0 kWh	Min 3.4 kWh	Min 8.2 kWh
	OR Services		Lighting of 1,000 lmhr/day	Electrical lighting, air circulation, television, and phone charging are possible			
2. Availability (Duration)	Hours per day		Min 4 hrs	Min 4 hrs	Min 8 hrs	Min 16 hrs	Min 23 hrs
	Hours per evening		Min 1 hr	Min 2 hrs	Min 3 hrs	Min 4 hrs	Min 4 hrs
3. Reliability						Max 14 disruptions per week	Max 3 disruptions per week of total duration <2 hrs
4. Quality						Voltage problems do not affect the use of desired appliances	
5. Affordability					Cost of a standard consumption package of 365 kWh/year <5% of household income		
6. Legality						Bill is paid to the utility, prepaid card seller, or authorized representative	
7. Health & Safety						Absence of past accidents and perception of high risk in the future	

Figure 1.4 ESMAP's Multi-tier Matrix for Access to Household Electricity Supply. The full report is available at: https://openknowledge.worldbank.org/bitstream/handle/10986/24368/Beyond0connect0d000technical0report.pdf?sequence = 1&isAllowed = y.

informed the approach of developers, practitioners, and researchers in addressing energy poverty.

These tiers should highlight that such attributes must be considered in the process of finding and designing ways to expand energy access. In many cases, design dimensions are informed by practical considerations. For example, in the development context, the provision of electricity is complicated by the challenges associated with building out the infrastructure for a centralized, main grid. Additionally, even where the grid is available, service delivery may be affected by frequent outages and poor service quality. In the context of high-intensity conflicts, there are additional concerns with investing in the capital-intensive activities associated with the centralized production and distribution of electricity, for which the given infrastructure may be particularly vulnerable to attack. For example, the high-intensity conflict in Syria saw damage to pipelines for oil and natural gas, as well as electricity transmission networks. According to the country's Minister of Electricity, "by early 2013, more than 30 of Syria's power stations were inactive and at least 40 percent of the country's high voltage lines had been attacked" (Gobat and Kostial, 2016). In Somalia—with the exception of some cities in Somaliland and Puntland—the private sector now largely owns and operates the energy supply system, because of massive disruptions to, and ultimately the incapacitation of, public energy infrastructure (AfDB, 2015). Although more research is needed at the intersection of energy, development, and conflict, Morris (2017) suggests that "a private sector-led approach focused on smaller-scale generation may in fact be more effective than alternatives in the face of violence."

As Morris suggests, the scale of energy provision may be an important consideration, and today's landscape and pace of technological innovation are such that alternative models to a centralized grid are increasingly possible. Decentralized (or distributed) electricity access options, including solar home systems, mini-grids, and storage technologies, open up new opportunities by offering greater flexibility and the potential for more localized control, often at lower costs. These solutions may be grid-connected, off-grid, or take a hybrid approach, and they may be deployed as alternatives to or enhancements of the traditional grid. For example, a solar home system with battery storage may be installed to provide electricity where a distribution network does not yet extend, or to provide electricity to supplement an unreliable grid connection. Decentralized options thus have the capacity to improve economic and social outcomes and, if renewables-based, offer the additional benefit of environmental gains.

Such options, of course, are not a panacea. As with many technologies, a mismatch between consumer desires and the capacity of the technology to meet such desires can create friction. This may mean that one system of energy provision is rejected or soon displaced in favor of another, or that people are unwilling to pay at levels needed to financially sustain a given system. Scholars and groups, including RAEL (see, in particular, Schnitzer et al., 2014), have highlighted a number of best practices in deployment, particularly for considering the use of microgrids to expand rural electrification. Recommendations cover multiple facets of the process, ranging from stakeholder engagement to the technical design of the system. For example, some best practices highlight the importance of engaging the community throughout the process of design and deployment; coordinating with government agencies (when possible, as some areas of the DRC qualify as a failed state) during project development; appropriately sizing the system's capacity to address current and future needs; promoting the use of electricity in particular support of commercial/productive loads; maintaining a system for appropriate and effective customer support and maintenance; providing accurate and transparent billing at affordable prices; and having zero tolerance policies for theft and sustained nonpayment (see also ARE, 2014; Brass, Carley, Maclean, & Baldwin, 2012; Palit and Chaurey, 2011; Sovacool, 2012). Among these, a broader, implicit recommendation is that an accurate understanding of the specific context for deployment is key, and failing to take these steps during project development can threaten the viability of or reduce the intended impact of the project. For example, in one study, attempts to provide electricity via solar microgrids in rural India were ultimately hindered by two major complications. The first was low demand for solar microgrids—largely due to beliefs that grid expansion and subsidized connections would soon arrive. The second complication involved issues with theft, and the unwillingness of local operators to punish people in their own communities for such actions (Fowlie, Khaitan, Wolfram, & Wolfson, 2016).

More generally, electrification activities are not inherently inclusive in providing potential benefits, given costs, sociopolitical factors, and governance approaches. For one, electricity by itself does little. Rather, the benefits of electricity depend substantially on whether complementary inputs are available or readily accessible—these may include inputs that directly consume electricity, such as electrical appliances, or more indirect inputs, such as roads, which can facilitate the transport of goods for or resulting from end uses of electricity. Furthermore, electrification may in fact serve to preserve inequities—in sub-Saharan Africa, for example, existing grids tend to favor wealthier communities, in turn reproducing disadvantages faced by communities without grid access (Avila, Carvallo, Shaw, & Kammen, 2017). Corruption and clientelism can contribute to these dynamics. In looking at electricity access in 46 sub-Saharan countries from 1990 to 2010, for example, Trotter (2016) found strong associations between democracy and increases in rural electrification—this led, in turn, to decreases in the rural–urban electrification divide. Other variables that correlated with increases in rural electrification included higher income per capita, national savings, and population densities (Trotter, 2016).

1.3 Outlook for energy for peace

An increasing body of literature assesses the microeconomic impacts of electrification in developing countries. While most point to overall improvements in wellbeing, the magnitude of impact varies. The interest of academics in better estimating and understanding the benefits of electrification on development outcomes provides opportunities for developers and academics to collaborate in project design, deployment, and assessment, to better understand the conditions under which electrification can promote positive socioeconomic changes, including outcomes relevant for gender equity and environmental benefits.

Given the developing and conflict context, energy providers should consider the potential future implications of any design for electricity provision. Jones and Howarth (2012; cited in Morris, 2017), for example, have warned against "creating future dependencies that could make fragile states very vulnerable to high energy prices." Other considerations might include designing systems with adequate flexible modularity to adapt to future needs, or an analysis of compatibility with future grid connections.

Importantly, for energy access projects to accurately address local needs, such projects should be developed in partnership with local communities, organizations, and businesses. In seeking to improve upon the existing state of electrification, developers must also consider the potential implications of displacing locally generated solutions, such as local owner operators of informal, diesel-powered distribution grids. Furthermore, an understanding of gender-related dynamics can provide insights into designing a system that accounts for, and more equitably supports, gender-related differences in desired uses of electricity.

To keep the discussion of the chapters to follow grounded, and to explore the role of energy in addressing crises and long-term development needs, this book focuses on diverse aspects of energy systems in peace building. Part I examines how energy policy has widened social divisions, taking the coal case of Kosovo and Vietnam. In Kosovo, lock-in of the massive usage of domestic coal has continuously damaged the quality of life, especially among the socially excluded people (Chapter 2). In Vietnam, the NIMBY problem of the location of coal-fired power plant has stirred the tensions between the central and the local. Both cases indicate an energy policy could possibly deepen the social divide (Chapter 3).

On the other hand, Part II explains a different type of energy provision can foster social inclusion. In Columbia, there are still a number of IDPs due to the past civil conflict (Chapter 4). The improvement of the electricity access contributes not only to their livelihood but also to the social inclusion. Understating this mechanism in the following case in Myanmar, electrifications in a certain context such as refugee camp should be treated as a trigger for the social inclusion as well as energy provision (Chapter 5).

Part III highlights the decision-making process in this context. Under a certain condition, energy provision could accelerate social inclusion as shown in the case in Columbia and Myanmar. Yet, critical view is given through the case in Bangladesh where Rohingya has freed from Myanmar. In the Rohingya camp, there is the politics of development discourse where the decision-making process is far from the refugees themselves (Chapter 6). As a consequence, the mechanism shown in Part II is in danger. On the other hand, the case in India shows the importance of the participatory decision-making to link the energy provision with social inclusions (Chapter 7). In the end, the South Sudan case strongly implies the critical importance of the nexus approach between renewable energy and peace building by highlighting its innovative financing mechanism called the Peace Renewable Energy Credit (Chapter 8).

References

AfDB (African Development Bank). (2015). *Somalia Energy Sector Needs Assessment and Action/Investment Programme.* African Development Bank: Tunis.

ARE (Alliance for Rural Electrification). (2014). Hybrid Mini-Grids for Rural Electrification - Lessons Learned. Alliance for Rural Electrification: Brussels, Belgium. Available at: https://ruralelec.org/publications/hybrid-mini-grids-rural-electrification-lessons-learned.

Avila, N., Carvallo, J.P., Shaw, B., & Kammen, D.M. (2017). The energy challenge in sub-Saharan Africa: A guide for advocates and policy makers. Oxfam Research Backgrounder. Available at: https://www.oxfamamerica.org/static/media/files/oxfam-RAEL-energySSA-pt1.pdf.

Brass, J. N., Carley, S., Maclean, L. M., & Baldwin, E. (2012). Power for development: A review of distributed generation projects in the developing world. *Annual Review of Environment and Resources, 37*(1), 107–136. Available from https://doi.org/10.1146/annurev-environ-051112-111930.

Crewe, E. (1997). Discourses of development: Anthropological perspectives. In R. D. Grillo, & R. L. Stirrat (Eds.), *Discourses of development* (pp. 59–80). Bloomsbury Academic.

E4SV. (2015). Sexual violence against women and girls: is energy part of the solution?. *Smart villages: new thinking for off-grid communities worldwide* [online]. Available at: http://e4sv.org/sexual-violencewomen-girls-energy-part-solution/.

ESMAP (Energy Sector Management Assistance Program), Multi-tier framework for energy access. Available at: https://www.esmap.org/node/55526. Accessed March 15, 2019.

Fleming, P. (2018). Tapping Renewable Energy for Peace in Somalia. *Sustainable Energy for All,* [online]. Available at: https://www.seforall.org/content/tapping-renewable-energy-peace-somalia.

Fowlie, M., Khaitan, Y., Wolfram, C., & D. Wolfson. (2016). Solar Microgrids and Remote Energy Access: How Weak Incentives Can Undermine Smart Technology. Working Paper. Available at: http://www.catherine-wolfram.com/uploads/8/2/2/7/82274768/microgrid-project.pdf.

Gobat, J. & Kostial, K. (2016). Syria's Conflict Economy. *International Monetary Fund Working Paper,* WP/16/123. Available at: https://www.imf.org/external/pubs/ft/wp/2016/wp16123.pdf.

Goldemberg, J. (1996). *Energy, environment and development.* London, England: Earthscan Publications.

Jacobson, A., Milman, A. D., & Kammen, D. M. (2005). Letting the (energy) Gini out of the bottle: Lorenz curves of cumulative electricity consumption and Gini coefficients as metrics of energy distribution and equity. *Energy Policy, 33*(14), 1825–1832.

Lee, K., Miguel, E., & Wolfram, C. (2017). Electrification and Economic Development: A Microeconomic Perspective. *EEG State-of-Knowledge Paper Series.* Oxford and Berkeley: Oxford Policy Management and the University of California, Berkeley.

Morris, R. (2017). EEG briefing note on energy, fragility and conflict. *Energy and Economic Growth: Applied Research Programme.* Available at: https://assets.publishing.service.gov.uk/media/5a26946f40f0b659d1fca8d5/Line_34_-_EEG_FCAS_Briefing_Note.28.06.2017.v1.pdf.

Moyo, D. (2009). *Dead aid: Why aid is not working and how there is a better way for Africa.* New York, New York: Farrar, Straus and Giroux.

Palit, D., & Chaurey, A. (2011). Off-grid rural electrification experiences from South Asia: Status and best practices. *Energy for Sustainable Development, 15* (3), 266–276. Available from https://doi.org/10.1016/j.eds.2011.07.004.

Schnitzer, D., Lounsbury, D.S., Carvallo, J.P., Deshmukh, R., Apt, J. & Kammen, D. M. (2014). Microgrids for rural electrification: A critical review of best practices based on seven case studies. United Nations Foundation. Available at: https://rael.berkeley.edu/wp-content/uploads/2015/04/MicrogridsReportEDS.pdf.

Scott, J. C. (1998). *Seeing like a state: How certain schemes to improve the human condition have failed.* New Haven, Connecticut: Yale University Press.
Sovacool, B. K. (2012). Deploying off-grid technology to eradicate energy poverty. *Science (New York, N.Y.), 338*(6103), 47–48. Available from https://doi.org/10.1126/science.1222307.
Stern, D.I., Burke, P.J., & Bruns, S.B. (2016). The impact of electricity on economic development: A macroeconomic perspective. *EEG State-of-Knowledge Paper Series.*
Trotter, P. A. (2016). Rural electrification, electrification inequality and democratic institutions in sub-Saharan Africa. *Energy for Sustainable Development, 34,* 111–129. Available from https://doi.org/10.1016/j.esd.2016.07.008.
World Bank. (2013). *Toward a sustainable energy future for all: Directions for the World Bank Group's Energy Sector.* World Bank: Washington, D.C. Available at: http://documents.worldbank.org/curated/en/745601468160524040/pdf/795970SST0SecM00box377380B00PUBLIC0.pdf.
World Bank. (2017). "DataBank: World Development Indicators," [online]. Available at: http://databank.worldbank.org/data/reports.aspx?source = world-development-indicators.

Kosovo's conflict coal: regional stability through coordinated investments in sustainable energy infrastructure

Noah Kittner[1,2] and Daniel M. Kammen[3,4]

[1]*Department of Environmental Sciences and Engineering, University of North Carolina at Chapel Hill, Chapel Hill, NC, United States* [2]*Department of City and Regional Planning, University of North Carolina, Chapel Hill, NC, United States* [3]*Energy and Resources Group, University of California, Berkeley, CA, United States* [4]*Goldman School of Public Policy, University of California, Berkeley, CA, United States*

Chapter outline

2.1 Introduction

One of the world's newest nations, the Republic of Kosovo, is at the forefront of the global debate on the future of coal financing.

Energy Policy for Peace. DOI: https://doi.org/10.1016/B978-0-12-817350-3.00004-3

After more than a decade of deliberations in the World Bank, the US Department of State, the European Union, and other multilateral development banks, donors, funders, and the government of Kosovo remain uncertain about the country's energy future. With the world's fifth largest supply of lignite coal, Kosovo continues to rely on coal for electricity, heating, and cooking, and this perpetuates the conflicts lingering from the wars of the 1990s. Regional stability in the Balkans hinges on future energy sector infrastructure investments that are poised to have an outsized impact on the environment and public health.

Emerging from decades of conflict, the countries in Southeastern Europe that had comprised the former Yugoslavia face unique energy challenges related to sustainability and reliability. In Kosovo, for example, negotiations surrounding the development of a new 500-MW coal-fired power plant have mired international financial institutions and regional governments in conflict and controversy. Continued reliance on lignite coal in Kosovo despite dramatic cost improvements in benign renewable energy technologies has the potential to significantly disrupt peace and stability in the region, owing to serious effects on air quality, state dependence on foreign lenders and resources, and civil unrest over high electricity bills and poverty. The various decision frameworks that have been employed and the regional actors that are considered set up the potential for a decade of conflict or peace through investments in electricity infrastructure. The sustainability, affordability, and reliability of energy sources will play critical roles in securing low-carbon, clean, and resilient infrastructure.

This chapter details the protracted and ongoing planning process around the proposed coal plant in Kosovo (Kosovo e Re) and highlights the role of regional cooperation and international attention to leverage sustainable energy projects and bring lasting peace and prosperity to Southeastern Europe. The planning challenges range from transitioning a society and economy highly dependent on lignite coal for electricity generation and household energy consumption to securing investment for a sustainable, equitable future in more distributed, local, and renewable resources.

The US government and the World Bank historically have played an outsized role in supporting and thus guiding the evolution of Kosovo's electricity infrastructure. This dates back to US involvement during Kosovo's transition to independence. However, a decades-old coal plant that urgently needs replacement could chart a new path forward in terms of available electricity infrastructure with the option to bring sustainable energy, improved reliability and affordability of resources, and new jobs to the region. This proposal largely centers on diverting investment for a new coal plant toward energy efficiency measures, distributed renewable energy projects, and transmission and distribution system retrofits to improve overall system planning. The pathways that have been outlined require coordination and cooperation with regional and domestic entities and would greatly reduce the risk of conflict over a large-scale centralized coal-fired power plant. The institutions in this case study remain key decision makers and are therefore implicated in Kosovo's future peace and prosperity.

The primary conflict risks presented by the current financial proposals include reliance on cooling tower water from Gazivoda Lake in Serbia, dependence on foreign capital, debt restructuring and nonpayment, public health risks from high levels of air pollution in population centers, and high retail rates for electricity. The lack of reliability and affordability from an option that also requires significant coordination with potentially hostile governments could threaten peace and stability in Kosovo due to power system investment.

As an alternative approach, this chapter describes the potential role of regional cooperation and international attention to reduce the conflict risks of building a new coal-fired power plant. Neighboring border countries, such as Albania and Serbia, and regional neighbors such as the European Union play critical roles in securing Kosovo's electricity future. For instance, greater integration of existing hydropower and transmission system assets in Albania could provide a strategic trading partner for balancing electricity markets and could improve neighborly relations between two countries that are interlinked closely by culture, outlook, and demographic profile. Furthermore, it would diversify Kosovo's electricity-trading partnership, since excess electricity generated in Kosovo could also meet Albania's seasonal shortfalls resulting from overreliance on hydropower. As Kosovo seeks accession to the European Union, improved public health, increased reliability, and improved energy efficiency measures in homes could all improve the quality of life and general welfare for a prosperous nation.

2.2 Regional conflict following the breakup of Yugoslavia and the history of Kosovo's electric power sector

In this chapter we discuss the emergence of the electric grid in the former Yugoslavia and explain the history behind the state-owned electric utility. We also show how the violent conflict in the 1990s between Serbia and Kosovo affects current grid infrastructure and power sector operations. Furthermore, we highlight the path dependence of coal for energy and its role in enabling Kosovo to become its own nation, separate from other Balkan countries. The ability to generate its own power independent of Serbia gave Kosovo electricity but saddled the newfound country with debt, lack of high-quality jobs, and inefficiencies of heating homes using lignite coal.

Foreign entities have played a significant role in determining Kosovo's electricity future. Even when Kosovo was a part of Yugoslavia in the 1960s, there were plans to utilize its rich lignite resources and form an electricity trading hub to the rest of Europe. Kosovo, with large lignite coal reserves, had historically been a resource hub for the rest of Yugoslavia. The coal-mining industry employed a large number of people, and the cold winters made coal a critical energy source. The construction of Kosovo A and B units allowed Kosovo to generate electricity and utilize domestic coal production.

The water crisis in Lake Gazivoda illustrates this in further detail. Lake Gazivoda provides the main source of water for cooling the existing Kosovo A and B power stations. The water rights and access are severely limited because most of the cooling water exists in territory controlled by Serbia. The strong dependence on water resources for providing power generation remains an underinvestigated topic, especially in transboundary issues when power purchase agreements are under negotiation. In 2014 one of the units in Kosovo A exploded, causing the plants to be shut down for repairs.

One could theorize that power dynamics in the Balkans enabled conflicts and exacerbated tensions. A lack of foreign involvement during the early days could have prevented some of the worst disasters in the Yugoslav wars. Perhaps vested interests in fossil fuels exacerbated the historical conflict in the region. Independence gave further recognition of coal as a national source of energy and necessary for Kosovo's prosperity. With independence and a lack of regional partnerships for funding new projects, Kosovo may have become more reliant on existing infrastructure, such as the coal-fired power plants for electricity generation. However, there have also been historically significant hydropower, biomass, solar, and wind resources in the country.

As a part of Yugoslavia, Kosovo may have had a wider ability to utilize more intermittent sources of electricity, including run-of-river hydropower, because of the transmission system network and infrastructure. However, after the war, the ensuing "Balkanization" of policies and politics led to continued development of the coal sector.

Foreign banks and lenders have enabled the privatization and total sales of a vertically integrated electric utility company to sell off distribution system assets. This represents a private sector takeover of the energy system. Multilateral development banks have the responsibility to act and lend in countries such as Kosovo, where the risk profile is higher than that for typical private sector investments.

The World Bank and other prominent groups have upheld the belief that Kosovo's vast resource of fossil energy is the only hope for emerging from poverty. However, there has been an underappreciation of the other renewable energy resources that are available in Kosovo that could also create a clean energy hub for the region. For example, the vast solar resources in Kosovo outpace those in Germany, and sustainably managed forest lands could provide high-grade biomass for alternative cofiring schemes or gasifiers.

2.3 Equity and residential energy poverty

Continued reliance on coal threatens social equity through high electricity costs for consumers, inefficient buildings, and dependence on coal for household heating. For many households, especially when some or all members are unemployed, electricity costs remain a significant monthly expenditure. Purchasing coal for heating adds to that cost and further burdens households with indoor air pollution from the coal ash, which contains toxic trace metals. Unfiltered combustion increases particulate matter emissions and exposure of women and children who spend more time in the house.

The toxic trace metals are not often discussed as a main factor to exacerbate social equity. However, the link is clear. Another main issue is that electricity through the existing transmission and distribution grid remains unreliable and the main source of electricity is coal. Wealthier households that seek to have a more reliable electricity supply may use a backup generator at their household or business. These generators could utilize renewable electricity or diesel fuel. However, coal is rarely used for backup generation. Even villages on the outskirts of urban areas have turned to small-scale hydropower as a way to generate more reliable electricity for the villages, despite the intermittency of small-scale hydropower.

Lignite coal is expensive and threatens the safety and security of families who depend on it for electricity and heating. If lignite coal cannot be purchased in one month, then the family may suffer from a lack of heat and hot water.

Kosovo suffers from high rates of electricity poverty. First, lignite coal is a low-quality fuel and contains toxic trace metals. Second, lignite coal is expensive for the average customer, and the high cost of the transmission and distribution system losses further increase costs borne by the average household. Third, unsustainable wood and coal collection for household heating increases asthma rates and respiratory issues while putting households in debt.

Winters in Kosovo are cold, and extreme cold events disproportionately affect poor households that lack proper heating infrastructure. This can contribute to increased hospital admissions, further burdening the household budget. For uninsured families, this can exacerbate poverty conditions. Most of the buildings in Kosovo lack insulation and have not been upgraded for more than 50 years. Therefore there is great need and great potential for energy efficiency improvement.

Building energy efficiency offers ways to improve insulation and conserve energy for household heating. Poor ventilation is common across the residential building sector. New electric heat pumps could alleviate some of these issues. Kosovo receives high amounts of solar radiation, which would enable a household solar + heat pump system integration.

High retail rates for electricity reduce economic and job opportunities in Pristina and the outlying areas.

Outdated building infrastructure and the continued reliance on coal exacerbate social equity issues through increased exposure to indoor air pollution and particulate matter, high monthly costs that are disproportionate to local incomes, and few chances for upward mobility to gain access to a healthier environment and increased income. By contrast, the energy efficiency sector in Germany and other parts of the European Union have created numerous job opportunities. Increased investment in energy efficiency and skills-retraining programs could offer a benefit to many who have lost their jobs since the Yugoslavia era wars, reduced employment at the coal mines, and lack of other economic stimulus.

Attention to the energy poverty issues and investment inequity is needed to deal with environmental justice issues. Kosovo has been alienated by neighboring countries and abandoned by the United States, and it struggles to keep up with other former Yugoslavian states. The majority Muslim nation has been marginalized since the time of the Ottoman Empire.

2.4 Dependence and lock-in around coal contributes to unemployment

Kosovo has a staggering rate of youth unemployment. The Kosovo Agency of Statistics reported a 57.9% rate of youth unemployment in January 2019. One of the major booming jobs creation sector in many economies is energy. However, studies have documented the relatively few jobs that fossil fuel–related industries can provide today. Developing the skills necessary to enable an energy efficiency and renewable energy–led economy could stimulate job growth, especially among youth. Furthermore, many young people who have job opportunities and a visa will seek employment in other parts of Europe where there is a thriving energy industry.

The dependence on coal prevents young entrepreneurs from accessing the investment capital they would need to start new businesses. Without a strong jobs sector there are few opportunities to accumulate the initial capital that is needed to start a business. Coal mining and electricity generation require few humans relative to their level of economic output. By contrast, energy efficiency programs offer a way to create new jobs that can benefit the economy.

Foreign entities continue to dominate the types of industries that are prevalent in Kosovo. Their control ranges from the investment requirements that are set forth by banks and lenders to the types of products that are sold in Kosovo. In addition, the water levels in Lake Gazivoda are controlled by Serbian entities. The large-scale energy infrastructure loans that have been offered thus far are mostly for new coal facilities or distribution system upgrades. However, making distributed energy resources and energy efficiency programs more affordable and viable from a lending perspective could be a strategy to reduce dependence on coal and water for cooling. The benefits would accrue not only by reducing air pollution, but also by stimulating employment, saving water, and providing a more secure electric power grid.

Kosovo often cites the fact that the world's fifth largest lignite reserves are located within the country's borders. However, this "resource curse" or dependence on coal ignores another energy resource that available for more than 5 hours per day: the sun. Kosovo receives more solar radiation than Germany, yet Germany has completely outpaced Kosovo in terms of solar photovoltaic (PV) installations. Furthermore, Germany serves as a great example of how increasing use of distributed rooftop solar PV technology can inject power into the distribution grid without causing significant technical challenges or overburdening people with costs. The current distribution grid in Kosovo suffers from technical and nontechnical losses due to theft of electricity and a lack of metering. By introducing more distributed solar technology and offering rooftop owners the opportunity to connect to the electric grid and sell the power they obtain, Kosovo could improve its distribution feeder system without having to invest in expensive load tap changers, voltage regulators, and frequency regulation devices. Additionally, Kosovo could improve the distribution system infrastructure and metering capabilities to monitor generation and consumption.

2.5 Debt and foreign capture

The proposed new coal infrastructure in Kosovo is likely to keep Kosovo in debt to other countries. For instance, the plans for a 500-MW coal-fired power plant unit to replace Kosovo A would be financed through a private loan.

Lending by the United States to Kosovo to finance coal capacity would place an undue burden on Kosovo to repay that debt. If Kosovo joins the European Union in the future and the coal-fired power plant becomes a stranded asset, there could be significant financial losses for the government of Kosovo resulting from this debt. If Kosovo eventually becomes a part of the European Union, developing key industries related to energy efficiency and renewable energy jobs would be an important path forward

Furthermore, dependence on coal has a high opportunity cost related to employment effects. The Republic of Kosovo remains one of the youngest countries demographically in Europe. However, the lack of jobs forces many young people to emigrate or take jobs that have significant occupational hazards. Mining of coal also takes a toll on respiratory illnesses and hospital admissions.

An expanded energy efficiency program for residential and commercial buildings could create more jobs per kilowatt-hour and enable and industrial transition for Kosovo. This includes new skills development and a reduction in the brain drain.

2.6 Household heating and conflict

Fuel wood is the dominant form of household heating energy in Kosovo and is a large source of environmental injustice, given that less efficient households suffer from a larger public health burden due to air pollution. A household survey found that around 70% of households use wood for heating in homes and 7% use lignite coal, which also presents environmental and health challenges. This presents an undue burden on many households to generate sufficient heat for their living spaces. Electric heat pumps offer an emerging alternative technology that, although it is still in its nascent phases, could provide significant health benefits to households using electricity. However, currently, electricity generation in Kosovo is highly inefficient and relies on lignite. Finding a balance between potential new technologies that provide heating services and the life cycle effects will be a critical measure for future viability.

Space heating is a seasonal issue. Very few homes or residences have conducted home energy assessments during the winter to assess the quality of structures such as walls or roofs that are needed for insulation and sealing. The cost of coal adds up in monthly increments, keeping many of the poorer households in poverty as a result of an expanded monthly expenditure for coal during the winter months to heat homes.

Energy efficiency retrofit programs could go a long way toward reducing the amount of coal that households use for heating. Improvements in window seals and better-quality insulation materials are necessary to reduce costs. Furthermore, doors often need to be replaced to reduce energy system losses. The efficiency of furnaces could be increased as well.

The Kosovo Agency of Statistics found that only 18% of households use electricity for heat in 2015. An expanded role for clean electricity to provide heat in buildings could help to reduce inequities in household emissions and reduce conflict with neighbors due to burning and collection of wood and coal.

2.7 Electric appliances and gender

Women bear the costs of much of the use of electricity, collect fuel wood and coal, and spend more time cooking indoors and taking care of children. Therefore they may be at higher risk of the health consequences of exposure to poor indoor air quality. Furthermore, the Kosovo Agency of Statistics reports differences in widespread electrical appliances that could contribute to overall national prosperity. Energy-efficient appliances may reduce the reliance on coal or other fuels for electricity, thus alleviating some of the regional challenges. Developing cross-border appliance trade with countries in the European Union would also improve the quality and availability of energy-efficient televisions, washing machines, hair dryers, irons, refrigerators, stoves, freezers, mixers, and microwave ovens.

Average temperature expected to rise as a result of to global warming, which will induce higher demand for access to air-conditioning in buildings during the summer months. As of 2015, air-conditioning units have only been present in approximately 1.6% of households. However, this number is expected to increase as temperatures rise and air-conditioning units become more widespread and affordable. Therefore increased access to air-conditioning could become a critical electrical load on buildings in the next decade. Better integration with EU markets for air-conditioning units and efficiency standards could help to avert peak demand instances of electricity during the summer that could lead to power outages or blackouts. The lack of consistent reliable electricity is also an issue that can lead to food spoilage and hospitalization of the elderly. These dramatic consequences of the lack of reliable electricity in households can damage appliances and exacerbate intrahousehold tensions between the common appliance users and the head of the household. Intrafamily conflicts due to energy could increase unless people have a reliable electricity supply that is matched with efficient appliances in homes.

2.8 Future electricity pathways

Based on studies conducted by Kittner, Dimco, Azemi, Tairyan, and Kammen (2016) and Kittner, Fadadu, Buckley, Schwarzman, and Kammen (2018), there are a number of reliable and cleaner pathways to reduce greenhouse gas emissions in the electricity sector while also significantly improving air quality, all at a lower direct cost basis as a result of the construction of new coal-fired power plants. Therefore Kosovo sits at an important crossroads in terms of future electricity infrastructure choices. These studies have concluded that there are significant advantages to meeting EU renewable energy targets and the industrial emissions directive before planning to join the European Union (Kammen & Kittner, 2015). Power integration and expanded power trade may be one of the key pathways toward diverting Kosovo from overreliance on coal and path dependency. However, owing to a lack of existing transmission infrastructure and constrained ability to import electricity from abroad, Kosovo continues on a coal-based path without external investment and deployment of alternative energy options.

Figs. 2.1–2.3 summarize the current state of the literature depicting future low-cost electricity pathways for Kosovo's power grid. One main issue is that if Kosovo e Re were to be built, there would be a large path dependency that could crowd out the employment and peace benefits of generating domestic electricity from sources such as solar and wind. Biomass electricity and hydropower along with strategic investments in transmission system infrastructure could open new power trading partners with Albania and other Balkan countries that are integrated with the

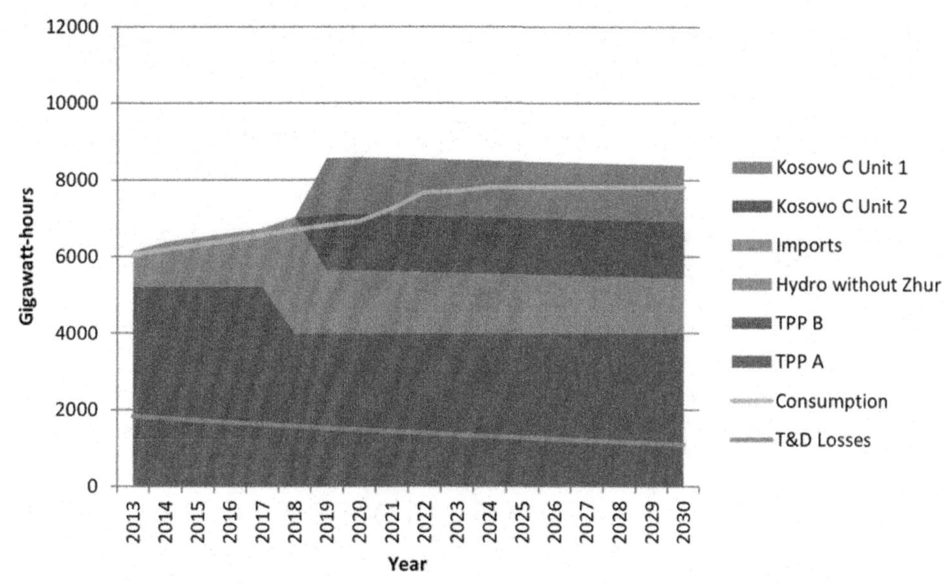

Figure 2.1 Projected sources of electricity generation in a future pathway that includes the development of a new 500-MW coal-fired power plant.

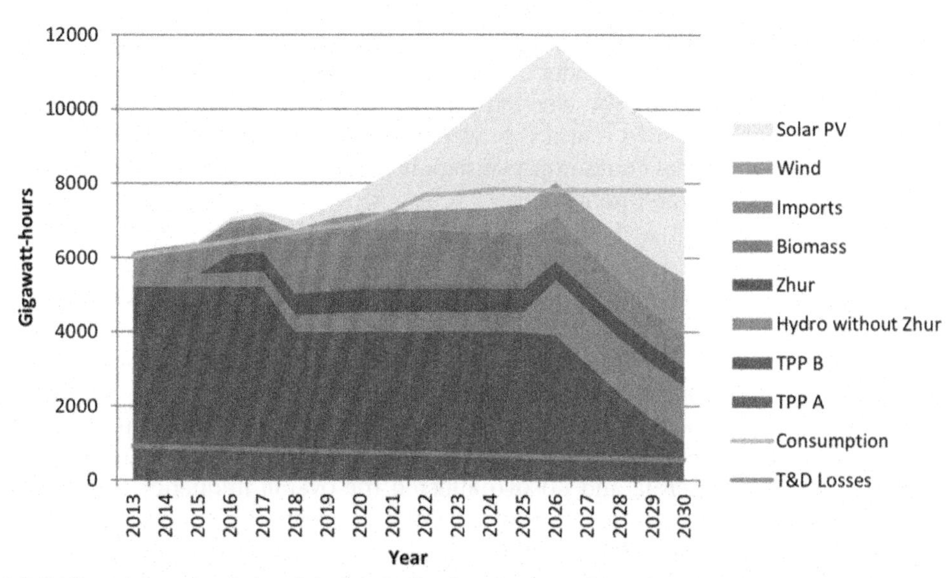

Figure 2.2 Projected sources of electricity generation in a future pathway that meets the EU 2030 targets for energy efficiency, CO_2 emission reductions, and renewable energy share. Adapted from Kittner, N., Fadadu, R. P., Buckley, H. L., Schwarzman, M. R., & Kammen, D. M. (2018). Trace metal content of coal exacerbates air-pollution-related health risks: The case of lignite coal in Kosovo. *Environmental Science & Technology, 52*(4), 2359–2367.

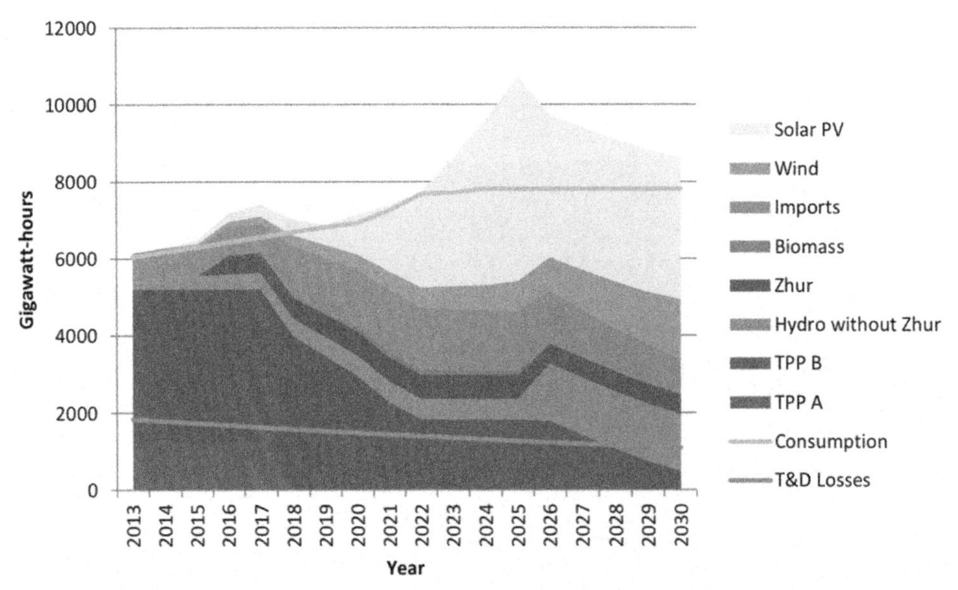

Figure 2.3 Projected sources of electricity generation in a future pathway that does not utilize natural gas as a source of power generation.

European electricity market. However, both biomass and small-scale hydropower introduce a number of contentious siting issues that could cause management conflicts over water resources and land to produce sustainable biomass outfits. Lignite coal has dominated the electricity generation picture since the construction of Kosovo A and B power stations. Weaning the power industry from coal could improve regional stability through shared infrastructure investments.

2.9 Air quality destabilizes regional security

Kosovo's power sector remains the single largest point source of pollution in continental Europe. Furthermore, frequent blackouts and inconsistent power generation scheduling affect the European power grid. For instance, in 2018 a dispute between Serbia and Kosovo affected the overall frequency in the European power grid when 113 GWh of electricity suddenly deviated. The operation of the power grid is a major security issue that requires cooperation between countries. Since many of the clocks rely on the electric grid frequency to keep time, there can be significant infrastructural consequences as a result of the lack of an integrated power market.

Transmission system investment plans could allow for better grid integration of electricity (Huang, Kittner, & Kammen, 2019). Regional links offer the opportunity to spread and diversify renewable electricity over a wider geographic region and smooth the power output. Additionally, strategic investments in the transmission system could reduce air pollution by reducing local generation of electricity and shifting the air pollution burden of electricity. Other operations could serve as a revenue stream for the export of electricity during times when demand is low.

Alternatively, Kosovo could suffer from grid isolation. As Europe develops more policies related to the industrial emissions directive and begins to count electrons from the source, Kosovo might find it more difficult in the future to buy and sell electricity in the European power market, especially if there is a high CO_2 emission intensity or high mercury or arsenic content in coal. There could be problems selling coal-fired electricity generation, Kosovo could be left with a stranded asset, and there could be further tensions and escalating legal battles related to the type of electricity being sold across state lines.

Better integration with Albania may offer a compromise to the issues discussed. Further economic and electric grid integration with Albania could be a first step to synergize economic development and regional security. Albania produces most of its electricity from hydropower rather than coal. There could be synergistic power grid operations of balancing increased solar production in Kosovo with existing grid infrastructure in Albania. The economic integration also ties into other sectoral goals and could increase the alliance in a region famous for heightened separation.

2.10 Investment conflict and the role of multilateral development banks in Kosovo's electricity sector

Finance plays an outsized role in Kosovo's reliance on coal. Future transitions to renewable electricity will require conflict resolution between foreign investors and contracting parties. The World Bank and other multilateral development banks or international financial institutions play a role in financing infrastructure. These institutions exert enormous influence on the infrastructure in the built environment and the direction of Kosovo's electricity sector. The cost of capital, cost of debt, and cost of equity could all cause future conflict without stakeholder engagement. Further dependence on coal may increase Kosovo's dependence on foreign loans, which could cause serious challenges in the future. The preference of multilateral development banks historically to finance fossil fuels exacerbates energy injustices and poverty. However, there are alternative paths for Kosovo's future, and the banks play a legitimate role in determining future outcomes. New opportunities have emerged with technological improvements for renewable electricity and energy efficiency to enable reliable access to electricity. Russian and Turkish streams of natural gas are currently not being discussed in a formal way but could introduce multiple geopolitical challenges that should be acknowledged and considered.

The lack of foreign capital investment promotes massive government debt and undermines Kosovo's sovereignty and autonomy. Therefore finding alternative funding sources and financial agreements that provide substantial investment in Kosovo and retain that investment in social and labor capital will be essential for developing and industrial transition in Kosovo and increasing renewable electricity generation. This could serve as an economic and sustainable development strategy while also reducing regional conflicts by empowering poor communities to move out of poverty.

Industrial policy could be the key to providing entrepreneurship in the energy sector, a sustainable environment, and new jobs. Multilateral development banks have the responsibility to invest in cleaner technologies and provide new ways for Kosovo to develop its own green banks for lending and development of new renewable energy projects and energy efficiency programs.

Significant investment will be required to transition away from dependence on coal in Kosovo. However, the direct costs are not as substantial as the costs of replacing existing infrastructure with coal. Further upgrades to the distribution and transmission system will go a long way toward improving reliability and cost recovery. Additional financing is important for leveraging other partners.

2.11 Conclusions

The attention toward building new coal plants in Kosovo shifts the focus toward future investment and development pathways. Continuing on a coal-based path may cause unanticipated conflicts related to water scarcity, air quality, and energy poverty. However, the main international financiers, including the World Bank, the US government, and the European Bank for Reconstruction and Development, have an opportunity to invest in energy efficiency and renewable energy options that could realign Kosovo's trajectory, stimulate the economy, and increase job opportunities.

The planning challenges to transition Kosovo away from coal could lead to positive cooperation in the region and across Europe in addition to promoting a more equitable and sustainable future. This chapter has reviewed the different critical areas that may cause conflict. However, peace building could emerge from shared infrastructure and coordinated investments. The chapter has highlighted the important role that multilateral development banks can play in accelerating a clean energy transition across the Balkans. It has highlighted the vast inequity that is present in Kosovo's electricity sector and has suggested that continued reliance on lignite coal will perpetuate slow violence and cause the persistence of inequality in the economy and the country.

References

Huang, Y. W., Kittner, N., & Kammen, D. M. (2019). ASEAN grid flexibility: Preparedness for grid integration of renewable energy. *Energy Policy, 128*, 711–726.

Kammen, D. M., & Kittner, N. (2015). Energy in the Balkans. *Economist (London, England: 1843), 411*, 8951.

Kittner, N., Dimco, H., Azemi, V., Tairyan, E., & Kammen, D. M. (2016). An analytic framework to assess future electricity options in Kosovo. *Environmental Research Letters, 11*(10), 104013.

Kittner, N., Fadadu, R. P., Buckley, H. L., Schwarzman, M. R., & Kammen, D. M. (2018). Trace metal content of coal exacerbates air-pollution-related health risks: The case of lignite coal in Kosovo. *Environmental Science & Technology, 52*(4), 2359–2367.

Further reading

Arifi, B., & Späth, P. (2018). Sleeping on coal: Trajectories of promoting and opposing a lignite-fired power plant in Kosovo. *Energy Research & Social Science, 41*, 118–127.

Imami, D., Lami, E., Zhllima, E., Gjonbalaj, M., & Pugh, G. (2018). Closer to election, more light: Electricity supply and elections in transition economies, the case of Kosovo. *Dubrovnik*.

Kosovo Agency of Statistics. *Energy consumption in households in 2015*. (2015). <https://ask.rks-gov.net/media/4309/energy-consumption-in-households-2015.pdf>.

Lappe-Osthege, T., & Andreas, J. J. (2017). Energy justice and the legacy of conflict: Assessing the Kosovo C thermal power plant project. *Energy Policy, 107*, 600–606.

Sáfián, F., Dóci, G., Csernus, D., Kelemen, Á., & Mao, X. (2019). Myths and facts about deploying renewables in the power systems of Southeast Europe.

3

Coal dilemma in Vietnam's power sector*

Nguyen Quy Tam[1] and Kensuke Yamaguchi[2]

[1]Fulbright School of Public Policy and Management, Fulbright University Vietnam, Vietnam [2]Graduate School of Public Policy, The University of Tokyo, Tokyo, Japan

Chapter outline

3.1 Introduction

3.1.1 Economic growth and power demand

Vietnam is a lower-middle-income country with a gross domestic product (GDP) per capita of $2563 in 2018 (World Bank, 2019). Since 1990, the country's GDP growth rate has averaged about 7% annually (see Fig. 3.1). Along with this strong economic growth and income improvement has come increasing electricity consumption, at 13.4% per year in the period of 2006–10, which slowed down to

*With special thanks to Professor David Dapice at the Vietnam Program, Harvard Kennedy School for his valuable inputs and comments.

Energy Policy for Peace. DOI: https://doi.org/10.1016/B978-0-12-817350-3.00006-7

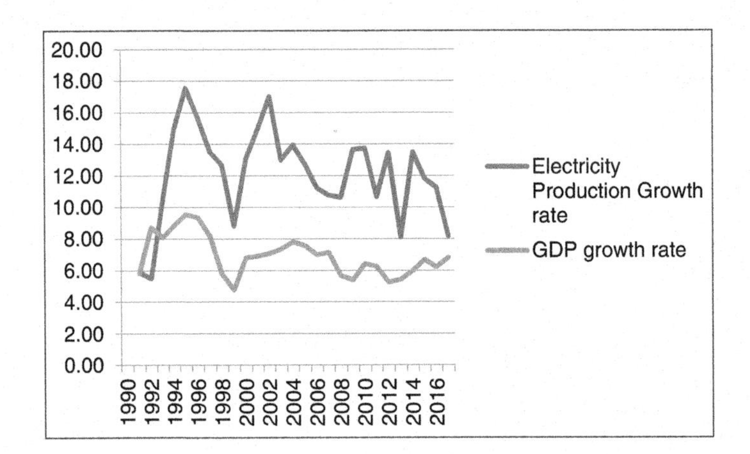

Figure 3.1 Annual growth of GDP and electricity production in Vietnam, 1990–2017. From GSO data.

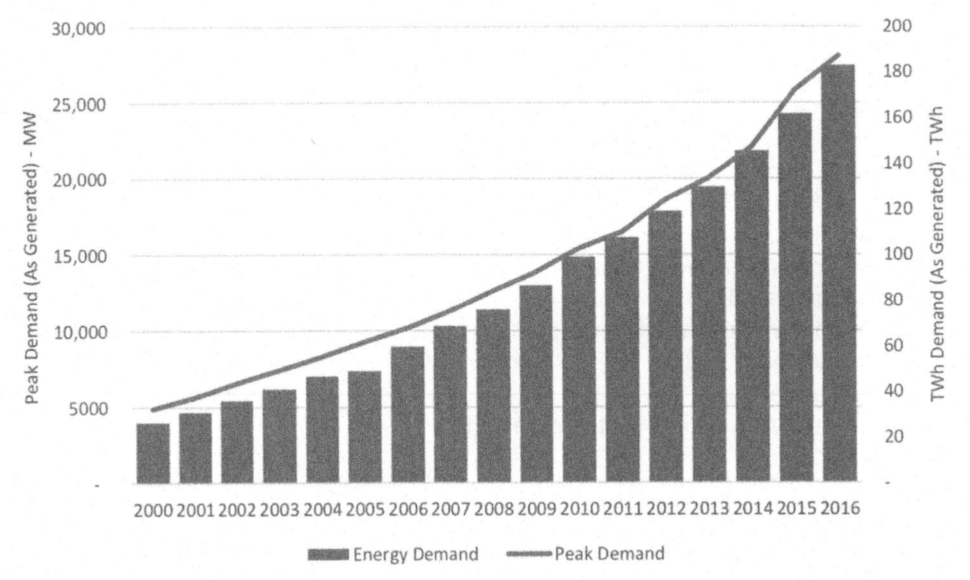

Figure 3.2 Energy and peak demand in Vietnam, 2000–16. From ADB (2017). TA-9012 VIE: Viet Nam Power Sector Reform Programme. Inception Report prepared by Intelligent Energy Systems and Energy Market Consulting Associates, December 2017.

10.3% per year from 2011 to 2015 (MOIT, 2017). On average, demand for electricity grew at a very rapid rate of 10%–11% during the 5 years ending with 2016 (ADB, 2017) and was projected to remain above the GDP growth for the next decade (see Fig. 3.2).

Fig. 3.3 shows levels of consumption in kilowatt-hours per capita in Vietnam and other countries in the region in 2015. Note that in 2015, Vietnam's GDP per capita in purchasing power parity (PPP) terms was less than half of China's, about one-third of Thailand's, and about one-fifth of Malaysia's. This reflects the high energy intensity of the Vietnamese economy,

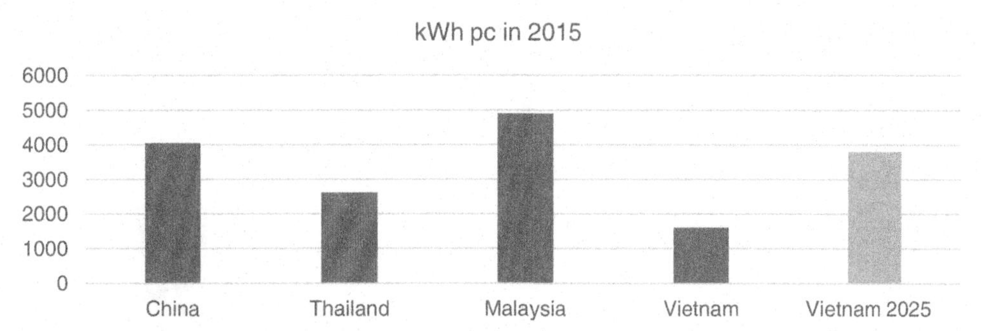

Figure 3.3 Per capita consumption in kilowatt-hours in 2015 for selected countries and Vietnam's projections for 2025 (David Dapice and Phu V. Le, 2017). From David Dapice and Phu V. Le (2017), Counting all of the costs: Choosing the right mix of electricity sources in Vietnam to 2025. Ash Center for Democratic Governance and Innovation.

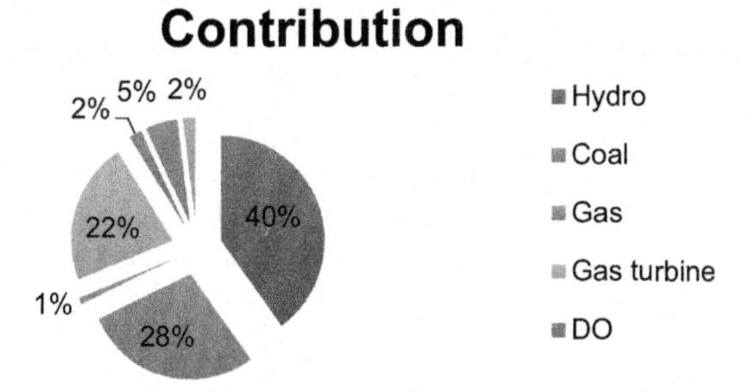

Figure 3.4 Electricity mix in Vietnam, 2014. From MOIT (2015). Presentation on Vietnam energy policy at IEEJ conference, August 2015.

which according to the World Bank's Energy Intensity Level of Primary database (2019), was 5.95, compared to China (6.69), Thailand (5.4), India (4.7), Malaysia (4.68), and Indonesia (3.52).

To meet this high intensity and increasing demand for energy, Vietnam has been successfully adding new electricity production capacity using different sources, with an average production growth rate of about 12% per year. By the end of 2014 the installed generation capacity of Vietnam reached 33.9 GW. Hydropower, coal-based thermal, and gas turbine accounted for the lion's share of the capacity mix at 40%, 28%, and 22%, respectively [Ministry of Industry and Trade (MOIT), April 2015] (see Fig. 3.4). Vietnam imported 1%–3% of its electricity from China (about 1000 MW) and from a hydropower project in Laos, the Xekaman 3 at a cost ranging from 6.68 to 7.02 cents per kilowatt-hour. (Vietnamfinance, 2019).

Even though the country's reserve margin of production increased dramatically from 2008 to 2011 and again to 2014, thanks to the commissioning of new thermal plants, the electricity supply in Vietnam still depends very much on hydropower, which has been developed mostly in the north and north central regions and is subject to seasonality (full operation only in the monsoon

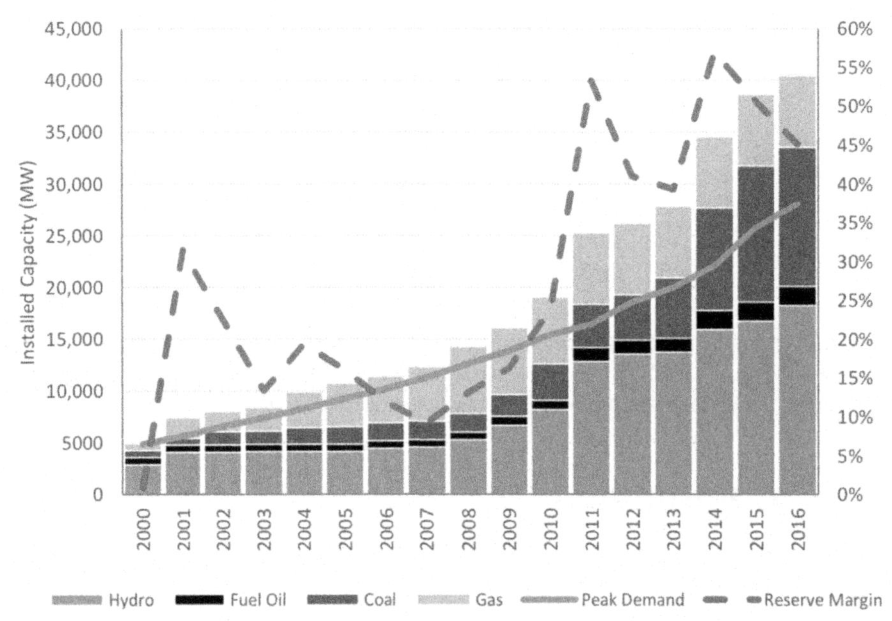

Figure 3.5 Installed capacity, peak demand, and reserve margin in Vietnam, 2000–16. From ADB (2017). TA-9012 VIE: Viet Nam Power Sector Reform Programme. Inception Report prepared by Intelligent Energy Systems and Energy Market Consulting Associates, December 2017.

season) (see Fig. 3.5). With limited transmission capacity, the electricity supply in the south—the country's economic powerhouse—can be tight when demand is high (ADB, 2017).

It is expected that by 2035, electricity demand will grow at an annual rate of 8% on average, which translates into need for an additional 93 GW of power generation capacity or double the current reserve capacity. Half of this capacity addition will come from coal-fired sources, and renewable energy will take up about 25% (MOIT, 2017).

Understanding the energy sector's key role in sustaining the country's growth momentum, the Government of Vietnam considers energy security its top priority, since a shortfall in energy supply can dampen economic growth. Nevertheless, the 2019 reserve margin is zero and shortages or rolling brownouts are becoming possible scenarios in the south (Vietnamplus, 2019).

3.2 Decision-making process of energy policy

3.2.1 Overview of power development plan 7

The key legislation that governs the future development of Vietnam's power sector is the National Master Plan for Power

Development. Vietnam's law on planning stipulates a 10-year span for each planning period with a vision for the next 10 years. In practice, there are interim reviews of plans (every 5 years) to update and accommodate new developments in the economy.

The master plan that is currently in effect is Power Development Plan 7 (PDP7) in the 2011–20 period, with considerations to 2030. The plan is quite comprehensive, covering the entire supply chain of the sector, from energy inputs to generation, transmission, and distribution. Among different sources of electricity, coal-fired power is expected to play a key role until 2030.

Approved by the Prime Minister's Decision 1208/Q-TTg in July 2011, PDP7's overall aim was to ensure energy security for the country by proposing a specific structure of power sources and new-generation capacity based on the projected GDP growth rate of about 7% annually to 2020, with considerations to 2030. While the plan indicated that clear priority would be given to the use of local sources of primary energy and the development of renewable energy, coal-fired power sources would account for almost half of both total installed capacity and production of electricity and would increase to more than half of the power sources by 2030 (Government of Vietnam, 2016), even though coal has to be imported. (See Tables 3.1 and 3.2).

At the Paris Climate Conference (COP21) in 2015, Vietnam committed to reducing greenhouse gas emissions by 8% by 2030 and up to 25% with international assistance. PDP7 was then revised and approved by the Prime Minister's decision 428/Q-TTg on March 2016, thereafter referred to as the Revised Power Development Plan 7 (RPDP7).

Renewable energy still receives priority for development, with special tax treatments and high FIT rates, and is adjusted

Table 3.1 Structure of power sources for installed capacity (Power development plan 7).

Source	2020	2030
	75,000 MW	146,800 MW
Hydropower (%)	25.5	15.7
Coal-fired thermal (%)	48	51
Gas-fired thermal	16.5% (2.6% liquid natural gas)	11.8% (4.1% LNG)
Renewable energy (%)	5.6	9.4
Nuclear power (%)	2.3	6.6
Imports (%)	3.1	4.9

Table 3.2 Structure of power sources in total production (Power development plan 7).

Sources	2020	2030
	330 billion kWh	695 billion kWh
Hydropower (%)	19.6	9.3
Coal-fired thermal (%)	46.8	56.4
Gas-fired thermal	24% (4% liquid natural gas)	14% (3.9% LNG)
Renewable energy (%)	4.5	6
Nuclear power (%)	2.1	10.1
Imports (%)	3	3.8

Source: Data from decision 1208/Q-TTg.

upward from 6% to 10,7% in the power mix for 2030.[1] Nevertheless, coal-fired power remains the major source of electricity production and is adjusted slightly downward to 42.6% from 49.3% in total installed capacity and to 53.2% from 56.4% in total production. As a result, coal power still accounts for about half of both total installed capacity and production (Government of Vietnam, 2016) (see Tables 3.3 and 3.4).

Although nuclear power was mentioned in both versions of the PDP7, it is no longer an option for development. By the end of 2016, the National Assembly of Vietnam had voted not to develop nuclear power, owing to several factors, including the increasing public debt burden, the need for nuclear waste treatment capacity, and other related risks. Both nuclear and coal are used for base loads, so coal may play an even larger role.

In brief, demand for both primary energy and electricity has grown strongly in Vietnam in the last two decades as the economy continues growing and becomes more energy intensive. Even though Vietnam has been successfully adding more generation capacity to meet this increasing demand, the risk of supply shortfall still exists, as the current supply structure still depends on hydropower that has been exploited most

[1]At the current rates of 8–10 cents per kilowatt-hour for renewable energy, there are tens of thousands of megawatts of proposed renewable energy investments that cannot be used as a result of grid constraints. Transmission costs are 1–2 cents per kilowatt-hour, while renewable costs in other countries are 3–5 cents per kilowatt-hour.

Table 3.3 Structure of power sources for installed capacity (Revised power development plan 7).

Sources	2020	2025	2030
	60,000 MW	96,500 MW	129,500 MW
Hydropower (%)	30.1	21.1	16.9
Coal-fired thermal (%)	42.7	49.3	42.6
Gas-fired thermal (%)	14.9	15.6	14.7
Renewable energy (%)	9.9	12.5	21
Nuclear power	—	—	3.6%
Imports (%)	2.4	1.5	1.2

Table 3.4 Structure of power sources in total production (Revised power development plan 7).

Sources	2020	2025	2030
	265 billion kWh	400 billion kWh	572 billion kWh
Hydropower	25.2%	17.4%	12.4%
Coal-fired thermal (%)	49.3	55	53.2
Gas-fired thermal (%)	16.6	19.1	16.8
Renewable energy (%)	6.5	6.9	10.7
Nuclear power	—	—	5.7%
Imports (%)	2.4	1.6	1.2

Source: Data from decision 428/Q-TTg.

feasible projects and is subject to seasonality, transmission limits, and climate change. With renewable energy still gaining ground but playing a small role in the current power mix, along with grid constraints, thermal power remains the major source of electricity supply, especially coal-fired power generation.

3.2.2 Planning process

According to Vietnam's Electricity Law, electricity planning is carried out at national and provincial levels. The MOIT is responsible for preparing the national plan in every 10 years, and the plan is subject to Prime Minister's approval. Provinces are charged with drafting a provincial electricity plan every 5 years, and these plans must be approved by MOIT to ensure conformity with the national plan. Thus the energy sector in Vietnam is highly centralized and operates under the state management of MOIT, which also takes charge of planning for coal, gas, and petroleum.

At the national level, the planning process starts with the Institute of Energy (IE) under MOIT. The IE drafts the power development plan, with inputs from Electricity of Vietnam (EVN), Petro Vietnam (PVN), and Vinacomin (TKV), the three state monopolies in electricity, oil and gas, and coal and mining, respectively. IE also consults with local provinces in regions where the electricity demand or base load is high or where the geographical locations are favorable to predetermined power sources for specific location for new generation capacity and for land planning.[2] These arrangements would be factored into the provincial plan for electricity. Having a billion-dollar investment project in their jurisdiction would presumably mean more GDP, job creation, and career advancement for provincial leaders. As a result, many economically underdeveloped provinces would hardly say no to the national plan.[3] The draft also includes a Strategic Environmental Assessment as required by the Ministry of Natural Resources and Environment before being submitted to MOIT for approval. MOIT will call upon the Directorate of Energy to review the draft. After rounds of adjustment and consultation, the Ministry will submit the draft to the Prime Minister for final approval.

When it comes to implementation, key investors (EVN, TKV, or PVN) will conduct detailed site planning under the auspices of MOIT. Local governments again play a crucial role in this phase when the location and technology of the power plant are to be determined. The Directorate of Energy will review the site plan for final approval by MOIT. Several other central ministries and local departments, such as defense, transport, construction, and people's committees of related provinces, are also involved in the process.

[2]New-generation capacity will be clustered in power centers that are defined as a cluster of two or more large-scale power plants fired by coal, gas, or diesel.
[3]At the local level, the provinces have their own power development plan for capacity of less than 50 MW.

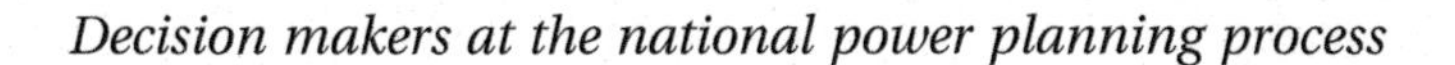

Decision makers at the national power planning process

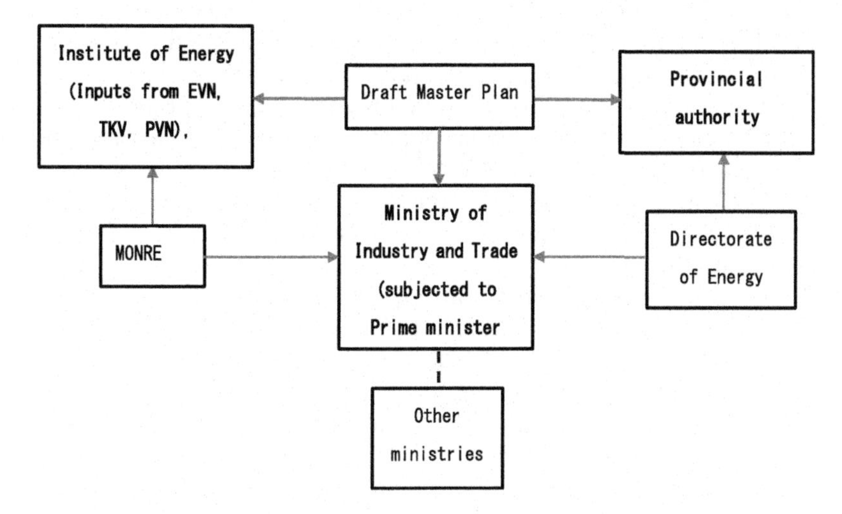

3.3 Project finance

3.3.1 Financial aspects of implementation

For the power sector in Vietnam, centralized planning of the entire system has helped to minimize the total cost of supply. The implementation of the plan is supervised by MOIT, and investment is often secured through direct involvement of EVN, PVN, and TKV. Recently, PPP projects, especially in power generation, have emerged as important additions to Vietnam's power sector capacity. However, the power grid and electricity transmission remain a state monopoly, whose capacity has not grown as fast as renewable production.

Traditionally, there have been three sources of finance for new-generation capacity in Vietnam's power sector: concessional lending from multilateral and bilateral financiers, borrowing from government and state-owned enterprises (SOEs), and local bank loans. These sources of finance have undergone some changes as a result of recent developments in both external and internal factors.

3.3.1.1 Relationships with international finance

First, the recent reversed trend of international finance to fossil fuel must be noted. OECD's stricter new rules on coal finance and international commitment to greenhouse gas emission reduction under the COP21 agreement have made international financing less available or accessible for coal-fired power projects (Paul Baruya, 2017). A report from Arabella (2016) shows that committed divestment from fossil fuel companies by organizations and individuals reached $5 trillion globally a year after the Paris agreement was adopted. In Vietnam an example following this trend is UK Export Finance's refusal to lend money to PVN's Long Phu 1 coal-fired power plant in Soc Trang Province because of the export credit agency's climate change priorities (NYT, 2018). This refusal to lend to any new thermal coal projects is driven by both environmental and financial concerns. Internationally, many existing coal plants are being idled as a result of competition from lower-cost sources.

In addition, Vietnam's national debt hit a new record of 63.7% of GDP at the end of 2016 and was slightly down to 61.3% by the end of 2017. In 2018 the government announced a borrowing plan of about $16.8 billion to pay its current debt; 72% of the amount comes from local loans, and the rest is from foreign sources (Vnexpress, 2018). The increasing public debt is made up mainly of government debt (overspending and government bonds), government guarantees (for borrowings from SOEs and state-owned banks), and provincial borrowings. The latest government guarantee in the power sector was given to the $1.6 billion Vinh Tan 4, for which 85% of the financing is from Korea Eximbank, Korea Trade Insurance Corporation, and Japan's JBIC.

The total value of government-guaranteed loans is estimated to be around $26 billion, 84% of which is foreign debt (Thanh Nguyen & Tuan Do, 2017). The government finds it increasingly hard to continue leveraging its own balance sheet and that of SOEs, especially for large-scale, capital-intensive infrastructure projects such as thermal power plants. This is one of the reasons why nuclear power was voted out of the new-generation mix. To deal with the rising public debt, in June 2016 the Ministry of Finance proposed that the government consider ending its guarantee to new projects starting from 2017.

3.3.1.2 Limited local finance

Lending to capital-intensive coal-fired power projects involves huge amounts of money that not many local banks can handle financially or technically. By regulation the credit line that a bank can extend to a single customer cannot exceed 15% of the bank's equity. The only window for local banks to do otherwise is Decision 1195 from 2005, which grants the State Bank of Vietnam authority to approve application from local credit agencies that can afford to lend to coal-fired power projects beyond the 15% threshold on a case-by-case basis (Government of Vietnam, 2005).

GreenID, a local nongovernmental organization (NGO) based in Hanoi recently released a report on sources of coal-fired power finance in Vietnam. The report found that of nine local banks that have financed coal power, three are state-owned commercial banks and one is a state-owned development bank. Together, they account for 92% of a total $2.5 billion of local finance sources (GreenID, 2017).

State-owned banks also engage in lending to other SOEs outside of the power sector. However, most of the SOEs are inefficient, which puts these banks under the stress of rising bad debts. For example, large state-owned corporations and general companies held 75%–80% of Vietnam Development Bank's outstanding debt in 2012. To a larger extent, SOEs accounted for about 70% of total bad debts in the entire banking system (Vnexpress, 2012).

About a quarter of the new-generation capacity projected in the RPDP7 by 2030 will be built by EVN and other SOEs, while the remaining finance will depend primarily on private sector investment (build, operate, and transfer - BOT or independent power producer - IPP) or other willing sources of international finance to realize the plan.

3.3.2 Chinese financing

3.3.2.1 Current status

China is neither a stranger nor a newcomer in terms of investment in and lending to Vietnam. By the end of 2017, total investments from China in Vietnam totaled $12 billion and ranked eighth (Saigontimes, 2017). Bilateral trade between China and Vietnam reached $93.69 billion in 2017, making China the country's largest trading partner (Industry & Trade News, 2018). As for official development assistance (ODA), there is no official data on Chinese ODA or concessional loans and export credits to Vietnam in general. Investment data of coal-fired plants in particular can be estimated only by collecting information on specific projects announced by related government agencies and project owners. GreenID, the local NGO that was cited previously, tried to build a database on sources of coal-fired power finance in Vietnam in 2016. It continued updating the database and recently released a report on its findings (GreenID, 2017). The database includes information that was collected from websites and official announcements of financiers, contractors, investors, and related ministries; from local and international media; and on all projects that are already in commission, are under construction, or will be built according to the national plan.

GreenID estimates that about $40 billion has been mobilized to build existing coal-fired power plants with a total installed capacity of 13,000 MW. To reach the planned capacity of about 55,000 MW, another $46 billion is required by 2030.[4] Of the invested $40 billion, 17% ($6.6 billion) came from domestic sources, including local equity and bank loans; about 52% ($21 billion) came from foreign sources; and the source of the remaining 31% ($12 billion) is unknown.

The foreign finance portion comprises $4.5 billion of equity, and about $16.5 billion of loans are from 23 international and national credit institutions. These include export credit agencies (52%), commercial banks (32%), and multilateral development banks (17%).

There are five key players in the ECA group: Exim Bank of China, Exim Bank of Korea, Korea Trade Insurance Corporation, JBIC of Japan, and Euler Hermes of France. China's Exim Bank accounts for more than half of the total foreign ECA finance and a fourth of the total foreign finance.

The commercial banks group includes five banks from China, five from Japan, and six from other countries such as Britain, France, Switzerland, Italy, and Singapore. Four large Chinese banks are China Development Bank, Industrial and Commercial Bank of China, Bank of China, and China Construction Bank. Altogether they account for 80% of total finance from this group and one-quarter of the total foreign finance.[5]

The multilateral development banks that still lend to coal-fired power projects in Vietnam are Japan's JICA and ADB with an estimate of $1.8 billion and $0.9 billion, respectively. Altogether, coal-fired power projects in Vietnam have been financed by nine countries, with China accounting for about 50% or $8.3 billion, Japan $3.7 billion, and Korea $3 billion.

There is uncertainty in GreenID's estimates of Chinese finance in Vietnam coal power projects beside the fact that 15% of the total finance sources cannot be identified. Still, their conclusion that the Chinese, Japanese, and Koreans have the leading roles in coal financing in Vietnam is supported by the global trend in coal financing by countries that the Natural Resources Defense Council and Oil Change International listed in their

[4]Note that the first 13,000 MW cost $3000/kW but the next 55,000 MW cost only $836/kW.

[5]US Export-Import Bank loans are for 20 years at an interest rate of 3.25%.

2016 report "Carbon Trap: How International Coal Finance Undermines the Paris Agreement" (NRDC, 2016).

The Carbon Trap report shows that from 2007 to 2015, a period that fits well into Vietnam's first PDP7, the four G20 nations of China, Japan, Germany, and Korea ranked at the top in international coal financing that totaled $62 billion. Vietnam was among the top three recipients with $10 billion; the other two top recipients were Indonesia ($11 billion) and South Africa ($7 billion). This trend will likely persist as Japan, China, and Korea all plan to continue financing future international coal projects totaling $20 billion.

Another report by the Climate Policy Initiative (CPI) in 2015 provides a much higher estimate of Chinese-supported overseas coal power projects from 2005 to 2014. During this period, China financed about $21–38 billion in projects and planned to pump another $35–72 billion into projects that were still in the planning stage. Again, Vietnam together with India and Indonesia received around 60% of the total Chinese overseas coal power capital (CPI, 2015).

A different estimate by Gallagher, K.P. confirms the rise of Chinese finance to coal especially in Asia. The estimate shows that Vietnam tops the list at $9.3b, followed by Indonesia ($8.6b), India ($7.7b) and Bangladesh (Gallagher, 2018) (S2.1b).

Remember that Vietnam's RPDP7 was approved in early 2016, with coal-fired power continuing to play a key role in new generation capacity to 2030, after the first PDP7 had already been partially implemented. The estimated $46 billion of capital requirement will probably maintain Vietnam's status as one of the top recipients of these internationally financed plans if the plans are fully carried out.

3.3.2.2 Reasons behind Chinese dominance

The CPI 2015 report seems to confirm what has happened in Vietnam when it comes to foreign financing of coal-fired projects by both public financing and private equity. Compared to Japan, whose financing motivations are mainly for the benefits of private firms that engage in coal power plant construction and equipment export, Chinese financing receives full support from the Chinese government to meet the country's domestic economic priorities and international influence ambition (CPI, 2015). These motives can be observed in infrastructure projects in Vietnam that use Chinese capital, particularly in the power sector, where Chinese financing is dominant via engineering, procurement and construction (EPC) contracts and emerging BOT investment.

Vinh Tan Power Center in South Central of Vietnam is a perfect illustrative example. Considered the most urgent project planned for development between 2013 and 2020, the center comprises four coal-fired power plants and is expected to generate and supply adequate electricity to the southern region and provinces. Of the four plants, three were awarded to Chinese EPC contractors: Guangdong Electric Power Design Institute (GEDI) and China Southern Power Grid for Vinh Tan 1, Shanghai Electric Group Company (SEC) for Vinh Tan 2, and China's Harbin Electric International Company for Vinh Tan 3. The EPC contract for Vinh Tan 4 was awarded to a consortium of Korea's Doosan Heavy Industries & Construction, Japan's Mitsubishi Corporation, and two local contractors (Sourcewatch, 2018).

The dominance of Chinese contractors in Vietnam's major projects is mirrored in other industries. In 2010, local news reported that Chinese companies "had won 90% of EPC contracts of Vietnam's major projects, most were in power, mining, petroleum, metallurgy and chemicals" (VCCI News, 2010). The Ministry of Planning and Investment (MPI) denied this estimate but acknowledged the relative dominance of Chinese EPC contractors in coal-fired power projects compared to those from other G7 countries. According to the Ministry's website muasamcong.mpi.gov (*public procurement* in English), by November 2010 a total of 118 EPC contracts in general had been awarded to contractors, of which 28 went to the Chinese (24%), 79 to locals (67%) and the rest to others. There were 16 coal-fired power projects among the 118 EPC contracts, and 12 of that (70%) were awarded to Chinese contractors by direct conferring (58%) and through competitive tendering (42%) (MPI, 2011). These figures do not include cases in which many local EPC winners subcontracted parts of their winning contracts to Chinese subcontractors.

There are two main reasons for the Chinese predominance in coal financing, according to the CPI 2015 report. First, like the loans of other international financiers, Chinese bank loans require a certain amount of Chinese involvement in the projects to which they lend. The 58% direct investment in projects in Vietnam is partly due to this lending requirement and partly due to the Vietnamese government's special decision known as the "duplex mechanism," the Decision 1195 mentioned earlier in the chapter, which is intended to speed up the construction of power projects classified as urgent.[6] Second, the Chinese

[6]Decision 1195 1195/Q-TTg dated November 9, 2005, stipulates that an EPC contractor of an ongoing project will be awarded another similar EPC contract if the contractor can arrange the required finance.

contractors are able to bid for EPC contracts at lower prices, given their comparative advantage in costs of construction, operation, and financing from home. This is the case for the other five projects mentioned on the MPI's website.

3.3.2.3 How competitive is Chinese financing?

Chinese financing is quite competitive indeed. A recent research by the Climate Policy Lab at the Center for International Environment & Resource Policy, The Fletcher School, Tufts University, which the author coauthored, found that the cost of Chinese financing of coal power plants in Vietnam ranges from 2.6% to 2.93% per year for 10 years plus the Libor rate.[7] This excludes all other fees, such as guarantee fee, financing arrangement fee, agency fee, insurance, and registration.

Besides the financial factor, Chinese contractors stand ready to meet all performance requirements, offering a quick disbursement process and implementation procedure. They are also good at forging close relationships with investors from the outset. These factors, combined with the pressing local need for project financing, weak environmental rule enforcement, and the reality of international financing being withdrawn from coal power projects, have given Chinese financing a favorable advantage in winning coal power projects in Vietnam.

In geopolitical terms, Vietnam is considered an important part of China's one belt one road (OBOR) initiative. Chinese infrastructure investment in Vietnam under this framework has increased recently. In the power sector, Vinh Tan 1 plant was the first Chinese BOT investment that has materialized under the OBOR.[8]

Power purchase agreement, government guarantee and undertaking, and coal supply agreement are critical to a PPP deal. Recently, a controversial case signals a possible new trend in Chinese financing of coal projects in Vietnam: No government guarantee is required. In July 2017, Geleximco, a local joint stock economic corporation, sought approval from the Prime Minister to build five thermal power plants, namely, Quynh Lap 1, Quynh Lap 2, Quang Trach 1, Quang Trach 2, and Hai Phong 3, under the form of PPP.[9] The proposed ownership structure is 20%–25% to designated SOEs, and the rest 75%–80% will be owned by the

[7]For comparison, US Export-Import Bank loans are for 3.25% per year for 20 years, and JBIC offerings range from 5.2% to 6.18% for 10 years in USD.
[8]"Vinh Tan 1 power plant largest Chinese investment in Vietnam": https://eng.yidaiyilu.gov.cn/home/rolling/20049.htm.
[9]Each plant will have two generation units with a total installed capacity of 1200 MW.

joint venture of Geleximco and Hong Kong United Investors Holding (HUI), a subsidiary of China's Sunshine Kaidi New Energy Group Co., Ltd. What is striking about this proposal, which stunned all electricity investors in Vietnam, is that the proposal seeks no government guarantee on the international finance portion of the project, no special treatment on project electricity outputs; and proposes to work with local governments to clear project sites within three months after receiving investment license as the main investor, a task that usually takes years to complete. The cost of one project is estimated about $2.1 billion and 80% of that will be backed by a group of six China's banks led by China Development Bank.

3.4 Choice of technology and environmental consequences

3.4.1 Choice of technology

The MOIT, the mastermind behind the national power plan, is charged with the responsibility of studying and promulgating regulations for types of technologies used and coal imported by coal-fired power plants to ensure compliance with environmental requirements and to reduce CO_2 emission.

There are specific environmental protection regulations that govern the thermal power industry. These regulations are promulgated by the Ministry of Natural Resources and Environment (MONRE) and the government, including MONRE's national technical regulations on emission of thermal power industry (QCVN 22:2009/BTNMT), wastewater (QCVN 40: 2011/BTNMT), and hazardous disposal (Circular 36/2015/TT-BTNMT). For solid waste and management of waste and residual disposal, dust, and cinders, Government's Decision 1696/Q-TTg and Decree 38/2015/N-CP will be applied. In general, environmental regulations that govern power sectors and other industries in Vietnam are quite comprehensive. However, compliance and enforcement are very different stories.

As for technology choice among coal plants, even though approval of a project is conditioned on whether it meets environmental requirements and is approved by MONRE, cost is the primary concern for investors. Given the high cost of advanced coal-burning technology such as supercritical (SC) and ultrasupercritical (USC), there have been only a handful of projects that use supercritical technology in Vietnam. These are Mong Duong 2, Vinh Tan 4 and the extension, Song Hau 1, and Vinh

Table 3.5 Emission parameters for a typical plant.

Emissions	Amount
SO_2	<300 mg/Nm^3
PM	<100 mg/Nm^3
NOx	<250 mg/Nm^3
CO	–
Hg	–

Tan 1 (all BOT projects). The technical explanation is that 5 years ago, SC and USC were expensive, and while efficiency improvement was about 3%–5% between subcritical and supercritical and about 10% between SC and USC, the cost difference was significant, as were the management and maintenance costs. The reapproval process was time consuming if there was any change in the project technical design. This explains the popularity of subcritical technology for most of projects that commenced before 2016–17 (about 21 projects.) Another factor is the lack of technical and managerial experience of local investors and consultants when it comes to working with SC and USC.

Table 3.5 shows emission parameters for a typical plant. The emission control system includes flue gas desulfurization using seawater, Electrostatic Precipitator System (ESP), and 210 m-chimney equipped with emission monitoring device. The plant produces 7.8 million metric tons of CO_2 per year. There is no collection or storage of CO_2. As required by regulation, all plants must have online emission-monitoring equipment with data being collected in real-time.

As for mercury, MONRE's circular 24/2017/TT-BTNMT requires measurement every 2 months. No automatic or continuous monitoring equipment is observed.

3.4.2 Environmental damage

According to a recent report by Vietnam Environment Administration (under MONRE), the 20 coal-fired power plants that existed in 2017 consumed about 40 million tons of coal per year, thus produced around 15.8 million tons of dust and cinder (85% fly ash and 15% cinder). About 25%–30% of the waste was reused as inputs for other industries, and the rest was stored or dumped in open pits that cover large areas and pose considerable risk to the surrounding environment. However, in 2017, only four of these plants met the local minimum environmental standards when checked by the Administration (Labor News, 2017). Also, Vietnam has been importing coal for domestic use since 2013. While coal imported from Indonesia and Australia, the country's two main suppliers, is lower in ash and sulfur compared to local sources, the scale of coal burning according to the power plan poses a considerable threat to the environment and public health. Indeed, according to a study by Koplitz, Jacob, Sulprizio, Myllyvirta, and Reid (2017), coal-related premature mortality in Vietnam by 2030 is projected to be 20,000 deaths if the current power development plan is carried out without a change in the power mix.

There have been a series of serious environment pollution incidents, mainly caused by, though not limited to, coal-fired power projects.[10] Since the commission of its first plant in early 2015, Vinh Tan thermal power center has continuously caused pollution in the neighborhood with its residual disposal. The peak of this pollution occurred in April 2015, when the National Highway 1 section in front of the power center was blocked for hours by thousands of local residents who were protesting against its pollution. This happened just a few months after the commercial inception of the second-generation unit.

The project has 64.7 hectares of open pits nearby to store its waste, from which coal dust and cinder were blown into residential areas by strong coastal wind (Vietnamnews, 2016). According to local news reports, each power unit of Vinh Tan 2 releases about 1500–2000 tons of dust and cinder into this pit daily. To deal with this ever-growing volume of waste, the project operators proposed dumping millions of tons of waste disposal into nearby marine-protected area. This proposal met with furious reactions and protests from local people and environment and marine experts.

A negative environmental impact can also be observed in Tra Vinh province, where the second major power generation center is located, comprising four plants: Duyen Hai 1, 2, and 3 and an extension of 3. Two of them went into operation in early 2016 even before their environment-related facilities were approved by environmental regulators. As a result, the livelihoods and living conditions of local residents have been seriously affected (Mekong Commons, 2014). Another kind of visible pollution caused by coal-fired power plants is water discharge at high temperature.

[10]The most serious case occurred when Formosa, a Chinese-financed steel complex in North Central Vietnam, dumped toxic liquid into the sea, killing aquatic life along hundreds of kilometers of the coastline.

The impact of hot water on local aquatic life has been recorded in the river near Quang Ninh thermal power plant, where fish and shrimp are killed by hot water or disposal discharged and leaked from the plant (VIR, 2018). Little information is available in the media about key emission factors of coal-fired power plants such as SO_2, PM, NOx, CO, and Hg. The reason might be that the emission cannot easily be physically observed by local residents or the media, unlike what happens with dust and cinder.

The MOIT in late 2016 named 27 projects sponsored by seven SOEs as being liable to cause environment pollution. The Ministry directed these SOEs to step up strengthening environmental protection measures in their projects and report progress accordingly (Vietnamnet, 2016).

There are different estimates on either the ratio or amount of coal dust and cinder being used as inputs to produce construction materials. Nevertheless, as much as 80% of waste from coal burning must be stored or dumped somewhere on a total estimated area of 127.4 hectares every year. This number will increase when the new generation capacity under RPDP7 is realized (Construction News, 2016).

There have been complaints and protests, some of which even turned violent, as in the case of Vinh Tan 2. However, pollution caused by coal-fired plants will persist regardless of sanctions imposed by authorities on these plants for their violations of environmental protection regulations. Plausible estimates of health and environmental costs of coal plants are in the range of 2–3 cents per kilowatt-hour, in addition to 7–8 cents in financial costs.

3.5 Coal dilemma: local veto against central push

As was shown in Section 3.2, the role of provincial governments is significant in Vietnam. And, concerning the cost and environmental aspects (Section 3.3), there are emerging cases of the reluctance of provinces to go for coal.

In 2006, Ninh Binh in the north of Vietnam was the first province to turn down a proposed coal-fired power plant and wanted it out of town. EVN, the investor of Ninh Binh 2 thermal plant, had worked out a scheme with JBIC on financing and of course with the local government on site selection. The province's reversed course came out when all preparations had been in place, including land clearance, earthworks, and construction facilities. Environment protection was the province's main reason. They already had Ninh Binh 1, a 30-year old coal-fired plant in town, and had experienced its impact on their environment. Also given the centrally approved plan to upgrade Ninh Binh town into a green cultural and tourism city by 2010, the leadership decided that another coal-fired power plant in town would not be desirable. The proposed alternative was to move the plant to an industrial park about 3 km out of town. As was mentioned earlier, approval of large-scale projects such as power plants is a lengthy and complicated process. Any change after approval would take it back almost to the beginning, with some economic costs to pay. Nevertheless, when pushed by the central delegation in a meeting in his province, the Chairman of Ninh Binh simply said, "We have no problem if you move the project to another province." Indeed, Ninh Binh 2 power plant was later replaced by Thai Binh 1 fire-power plant in Thai Binh Province, which is about 60 km to the northeast of Ninh Binh.

The second case took place in Bac Lieu province.[11] Under the RPDP7, Bac Lieu would be home to a new power center with two coal-fired power plants, commencing in 2030, with a total installed capacity of 1200 MW. The province is also an aquaculture hub that ranks second in terms of shrimp output nationwide. Aware of the environmental consequences of coal-fired power plants, especially after the experience of Vinh Tan power center, in 2016 the Bac Lieu leadership made a decisive move to withdraw from the national power plan. The province gave a politically judicious justification that they needed to fulfill the task that the politburo had assigned them to become a national champion in shrimp production, which demands a coal-free environment. The province later succeeded in attracting a 4 billion USD liquid natural gas (LNG)–fired power plant as an alternative to coal and several wind power investments.

Tien Giang, another province in the Mekong Delta, about 52 km from Ho Chi Minh City, is the next case. The RPDP7 in 2016 designated Tien Giang to be the home of Tan Phuoc thermal power center at a total installed capacity of 2400 MW, to commence in 2027. Later in 2016, the project investor EVN met with Tien Giang leadership to discuss implementation of Tan Phuoc power center. The actual plan was revealed to be a complex of four coal-fired power plants with a capacity of 1000 MW each. This power center would require an area of 420 hectares and use imported coal via waterway transportation. In response, the province refused, reasoning that the original plan

[11]In the Mekong Delta there will be seven power centers that aim to provide about 36,000 MW to the entire southern region.

mentioned only thermal plants, which the province would welcome if LNG were the energy of choice, not coal. The environment was again the main justification. A negotiation process then took place with a series of meetings at which the state investor was trying to push for coal and the provincial leadership was standing firm. The debate seemed to end when the province finally issued an official document in August 2018 to EVN in which the province asserted its sole support for an LNG thermal power center and carefully cited the Provincial Communist Party committee's strong endorsement of this decision.

The fourth and most recent case happened in Long An province, also in the south of Vietnam. RPDP7 assigned two plants in Long An with a total installed capacity of 2800 MW by 2027. Two locations were chosen through the location-planning process with one in close proximity to Ho Chi Minh City. Immediately, Ho Chi Minh City raised its voice.[12] The city claimed that inevitable pollution from the coal-fired plant would affect its major container port and the nearby new urban area located downwind of the proposed plant location. Another reason that was given was that the plant would inevitably cause damage to Can Gio mangrove forest, a UNESCO biosphere reserve and the so-called city's lung. Facing objection from Ho Chi Minh City and after witnessing environmental problems that Binh Thuan has experienced,[13] the Long An administration officially asked MOIT to change the proposed fuel types and technology of the plants, requiring that LNG be the fuel of choice. This came after the administration had first agreed to use Japanese or Korean coal-burning USC technology for both plants. Then the administration required LNG for the plant near Ho Chi Minh City and now would not accept anything other than LNG. The request was turned down by MOIT, and in effect no coal-fired plant will be built in Long An.

3.6 Conclusions

The chapter has provided a narrative on recent development of the energy sector in Vietnam, in which coal has been a primary incremental source of energy. As in some other developing

[12]Ho Chi Minh City is the economic powerhouse of Vietnam with significant contribution to national budget.

[13]By design, Vinh Tan 2 power plant has a pit of 38.37 hectares, with a capacity of 9.3 million cubic meters of ash and cinder and a maximum height of 27 m. Since it opened in 2015, it has released about 3.9 million cubic meters of ash and cinder, mounting up to 16 m high.

countries, energy policy in Vietnam leans toward coal as the fastest and controversially cheapest way to ensure energy security and economic development. Although the country has been successful in mobilizing resources to produce greater electricity output, resulting admirably in the electrification of almost 100% of the population, the low-hanging fruit has been picked and the road ahead will be bumpy, especially when coal continues to be chosen as the main source of electricity. According to MOIT, of the 62 power projects with capacity of more than 200 MW each, only 15 are on schedule. The remaining 47 are either behind schedule or of unknown schedule with regard to RPDP7 (Vietnamnet, 2019). Most of them are thermal power plants. Among several reasons that have led to these delays, one can see two critical factors emerging as patterns in preventing the country from using a pro-coal policy to find the much-needed energy and ensure social stability for economic development. They are constraints to coal financing and local resistance to coal.

There has been a global tendency to move away from coal as it is considered a dirty source of energy with serious health effects and global environmental consequences. Coal is also proving to have a higher financial cost than renewable energy or combined-cycle gas in many countries. After the Paris Agreement on Climate Change, international financing institutions have stopped financing coal, or if they still do, they require much stricter environmental compliance. This magnifies the already strained market for project financing in Vietnam. There are only a few lenders left for coal projects, namely, Japan, Korea, and China, which have emerged to gain significance in absolute dollar terms. However, Japan and Korea are facing criticism from the international community for breaching their commitment against coal, and obvious environmental pollution caused by Chinese-financed coal-fired power plants in Vietnam has caught serious attention from both the general public and environmentalists locally. This gives rise to the second defense against coal power development in Vietnam.

While local NGOs, the media, academia, environmentalists, and residents, especially those living near power centers, do raise concerns and even protest against pollution caused by coal-fired power plants, the most effective line of defense against coal lies unexpectedly in the hands of local provincial governments. Being key stakeholders in the national and local electricity planning process, they in effect hold critical and implicit power in vetoing the pro-coal policy. Some of them have done so tactically on the grounds of environmental concern and political mandate to stop coal-fired power plants from being built on their turf. (Often, the plants mainly provide power for major cities but are placed in poor, rural areas.) Of course, there are prices to pay for taking an environmental stand, such as unrealized economic development potential provided by coal plants, including job creation, increased revenue, and political support that promises advancement for local leaders. However, environmental costs caused by coal-fired power plants are happening in other provinces, where their negative effects on economic activities such as agriculture and the social unrest they cause, are becoming too much to bear for local governments. The current national power plan designated six provinces in the Mekong Delta in the south of Vietnam to host 14 new coal-fired plants totaling 18 GW. Three of the provinces (that is, half) have effectively refused to accommodate any coal-burning activities in their territory, all basing their decisions on environmental concerns.

For this reason, even if financing is resolved, the Revised Power Development Plan 7 and the coming Plan 8 will continue to fall short of meeting their targets and thus the economic development goals if coal continues to be planned as the main source of energy. Fortunately, the costs of renewable energy are lower than the cost coal in most countries, and if the grid were upgraded, the energy mix could be weighted much more heavily toward cleaner energy, with lower economic and health costs. There are also signs of greater willingness to use LNG as an energy source, so the future role for coal may well diminish.

References

ADB. (2017). *TA-9012 VIE: Viet Nam Power Sector Reform Programme.* Inception Report prepared by Intelligent Energy Systems and Energy Market Consulting Associates, December 2017.

Arabella. (2016). *The Global Fossil Fuel Divestment and Clean Energy Investment Movement.*

Construction News. Using coal burnt residual as construction materials. Ministry of *construction's news piece.* (2016). <http://www.baoxaydung.com.vn/news/vn/vat-lieu/su-dung-tro-xi-lam-vlxd-tiet-kiem-tai-nguyen-than-thien-moi-truong.html>.

CPI. Slowing the growth of coal power outside China: The role of Chinese finance. Authored by Morgan and Wang. (2015). <https://climatepolicyinitiative.org/publication/slowing-the-growth-of-coal-power-outside-china-the-role-of-chinese-finance/>.

David Dapice., & Phu V. Le. (2017). *Counting all of the costs: Cho*osing the right mix of electricity s*ources in Vietnam to 2025.* Ash Center for Democratic Governance and Innovation.

Gallagher, K. P. (2018). *China's* global energy finance database. Global Economic Governance Initiative, Boston University.

Government of Vietnam. *Decision 1195 on special mechanisms and policies granted to urgent power projects.* (2005). <https://thuvienphapluat.vn/van-ban/Dau-tu/Quyet-dinh-1195-QD-TTg-co-che-chinh-sach-dac-thu-de-thuc-hien/108256/noi-dung.aspx> Accessed 15.05.18.

Government of Vietnam. *Decision 428/QD-TTg on Revised National Power Development Master Plan for the 2011–2020 period with vision to 2030.* (English version translated by GIZ). (2016). <http://gizenergy.org.vn/media/app/media/PDF-Docs/Legal-Documents/PDP%207%20re.vised%20Decision%20428-QD-TTg%20dated%2018%20March%202016-ENG.pdf> Accessed 30.03.18.

GreenID. Coal-fired *power plants dev*elopment: A *project finance perse*pctive. (in Vietnamese). (2017). <http://www.greenidvietnam.org.vn/view-document/59b638d4a5d814d0281c8743> Accessed 13.04.18.

Industry and Trade News. *Vietnam-China trade might reach $100 billion in 2018.* (2018). <http://baocongthuong.com.vn/nam-2018-thuong-mai-viet-nam-trung-quoc-co-the-dat-100-ty-$.html> Accessed 25.05.18.

Koplitz, S. N., Jacob, D. J., Sulprizio, M. P., Myllyvirta, L., & Reid, C. (2017). Burden of disease from rising coal-fired power plant emissions in southeast Asia. *Environmental Science and Technology, 51,* 1467–1476.

Labor News. *Several thermal power plants pollute environment.* (2017). <https://laodong.vn/kinh-te/van-con-nhieu-nha-may-nhiet-dien-gay-o-nhiem-moi-truong-515514.ldo> Accessed 26.05.18.

Mekong Commons (2014). <http://www.mekongcommons.org/three-massive-coal-power-plants-bring-misery-ecological-risks-local-people-vietnams-mekong-delta/> Accessed 30.05.

MOIT (2015). *Presentation on Vietnam energy policy at IEEJ conference,* August 2015.

MOIT. *Vietnam energy outlook report.* Danish Energy Agency. (2017). <https://ens.dk/sites/ens.dk/files/Globalcooperation/Official_docs/Vietnam/vietnam-energy-outlook-report-2017-eng.pdf> Accessed 11.04.18.

MPI. *The truth about exessive Chinese EPC in Vietnam.* (2011). <http://muasamcong.mpi.gov.vn:8082/NEWS/EP_COJ_NEW005.jsp?newsId = 82> Accessed 30.04.18.

NRDC. (2016). Carbon trap: How international coal finance undermines the Paris agreement. Available from https://www.nrdc.org/sites/default/files/carbon-trap-international-coal-finance-report.pdf.

NYT. *The U.S. may back a Vietnam coal plant. Russia is already helping.* (2018). <https://www.nytimes.com/2018/01/26/business/exim-bank-vietnam-russia-coal.html> Accessed 19.05.18.

Paul Baruya (2017). International finance for coal-fired power plants. IEA Clean Coal Center, April 2017, pp. 20–21.

Saigontimes. Surge in Chinese investment in Vietnam. (2017). <https://www.thesaigontimes.vn/265946/Dau-tu-Trung-Quoc-vao-Viet-Nam-tang-manh.html> Accessed 25.05.18.

Source Watch. *Vinh Tan power station.* (2018). <https://www.sourcewatch.org/index.php/Vinh_Tan_power_station>.

Thanh Nguyen & Tuan Do (2017). *Diagnosing Vietnam public debt.* A prepared paper for Vietnam Economic Forum 2016.

VCCI News. *Chinese firms win 90% of Vietnam EPC contracts.* (2010). <http://vccinews.csb.vn/news/21177/.html> Accessed 25.05.18.

Vietnam Finance. *Vietnam buys electricity from Laos and China, also sells to Cambodia.* (2019). <https://vietnamfinance.vn/viet-nam-dang-mua-dien-tu-lao-trung-quoc-va-ban-dien-cho-campuchia-nhu-the-nao-20180504224226373.htm>.

Vietnamnet. (2016). <http://english.vietnamnet.vn/fms/environment/165592/gov-t-names-27-potential-polluters.html>.

Vietnamnet. *47 power projects behind schedule, risking power shortage.* (2019). <https://vietnamnet.vn/vn/kinh-doanh/dau-tu/47-dai-du-an-dien-cham-tien-do-canh-bao-nguy-co-thieu-dien-cao-542831.html> Accessed 02.08.19.

Vietnamnews. *Vinh Tan power plant still polluting residantial area.* (2016). <https://vietnamnews.vn/environment/281921/vinh-tan-power-plant-still-polluting-residential-areas.html#s4UiYmwjQMhxzp8o.97> Accessed 01.02.16.

Vietnamplus (2019). <https://en.vietnamplus.vn/power-import-from-china-rises-by-167-in-first-5-months/155401.vnp>.

VIR. *Suffering downstream of polluting coal plants.* (2018). <http://www.vir.com.vn/apicenter@/print_article&i = 57050> Accessed 29.05.19.

Vnexpress. *Bad debt help by SOE reached VND 200 trillion.* (2012). <https://kinhdoanh.vnexpress.net/tin-tuc/vi-mo/no-xau-cua-doanh-nghiep-nha-nuoc-toi-200-000-ty-dong-2722865.html> Accessed 25.04.18.

Vnexpress. *Vietnam plans to spend $11.3 bln to repay government debt in 2018.* (2018). <https://e.vnexpress.net/news/business/vietnam-plans-to-spend-11-3-bln-to-repay-government-debt-in-2018-3741021.html> Accessed 25.04.18.

World Bank. *World development indicators.* (2019). <https://data.worldbank.org/indicator/NY.GDP.PCAP.CD?locations = VN> Accessed 26.07.19.

World Bank. (2019). *Energy intensity level of primary energy.*

Further reading

Customsnews. *The ten biggest trading partners of vietnam in 2018.* (2019). <https://customsnews.vn/10-biggest-trade-partners-of-vietnam-in-2018-9639.html> Accessed 24.01.19.

EVN. *Vietnam electricity annual report.* (2017). <https://evn.com.vn/userfile/files/2017/EVNAnnualReport2017-web.pdf> Accessed 29.09.18.

EVN news. *Majority of power projects delayed.* (In Vietnamese). (2019). <https://www.evn.com.vn/d6/news/Phan-lon-cac-cong-trinh-nhiet-dien-cham-tien-do-6-13-23288.aspx> Accessed 14.03.19.

EVN news. *Electricity generation prices of coal-fired thermal power plants.* (2019). <https://en.evn.com.vn/d6/news/Electricity-generation-prices-of-coal-fired-thermal-power-plants-not-exceeding-VND-15687-kWh-66-163-480.aspx> Accessed 23.01.19.

Government of Vietnam (2009). *National technical regulation on emission of thermal power industry, QCVN 22:2009/BTNMT.*

Herve-Mugnucci and Wang. *Slowing the growth of coal power outside China,* Climate policy initiative. (2015). <https://www.adb.org/projects/49196-003/main#project-pds>.

Reference site for legal documents. (Library of laws). <http://www.thuvienphapluat.vn>.

Tuoitrenews. *Megastory: No more land for coal burned diposals in 2–3 years.* (in Vietnamese). (2018). <https://tuoitre.vn/se-khong-con-cho-chua-tro-xi-cua-nhiet-dien-than-20181223013753997.htm> Accessed 23.12.18.

Vietcombank Securities (2016). *Vietnam power industry.*

Vietnam Energy online. *No more reserves in electricity generation.* (in Vietnamese). (2019). <http://nangluongvietnam.vn/news/vn/dien-luc-viet-nam/he-thong-dien-dang-van-hanh-trong-tinh-trang-khong-co-du-phong.html?fbclid = IwAR0C0Hev0EVNNxrBW2sUQ0b3F3KNFNWyz_UGkp7oX8PHueqz1_gY3lLYpzY> Accessed 29.05.19.

World Bank (March 2018). *Vietnam Maximizing Finance for Development, Energy Infrastructure Assessment Program* (VN Energy Infra-SAP).

4

Energy access and durable solutions for internally displaced people: an exploration in Colombia

Daniel Del Barrio-Alvarez[1], Jairo Alberto Garcia-Riveros[2], Jose Luis Wong-Villanueva[3] and Kensuke Yamaguchi[4]

[1]Department of Civil Engineering, The University of Tokyo, Tokyo, Japan [2]Social Innovation Science Park, Corporación Universitaria Minuto de Dios - UNIMINUTO, Colombia [3]Department of Urban Engineering, School of Engineering, The University of Tokyo, Tokyo, Japan [4]Graduate School of Public Policy, The University of Tokyo, Tokyo, Japan

Chapter outline

Energy Policy for Peace. DOI: https://doi.org/10.1016/B978-0-12-817350-3.00003-1

4.1 Introduction: realizing durable solutions for internally displaced persons and the role of energy projects after the conflict

Forced displacement in one of the most immediate consequences of armed conflict. The global displaced population is rising at a worrisome rate. In 2018 the UNHCR reported 70.8 million forcibly displaced people worldwide (UNHCR, 2018). Political instability, economic necessity, and climate change continue to be factors in the increase in forced displacement. Of the total number, nearly 41.3 million are internally displaced people (IDPs), 25.9 million are refugees, and 3.5 are asylum seekers. IDPs are a special group of vulnerable forcibly displaced people because they are not entitled to protection from the international community, as they have not crossed the national borders. Furthermore, their vulnerability might be diffused among other vulnerable groups within the country, thus increasing their invisibility.

The international framework for durable solutions for IDPs is oriented toward the effective removal of vulnerabilities caused by the displacement (Kalin, 2005). Eight criteria determine the extent to which a durable solution is achieved. Granting safety and security is the first condition. It must be note that the removal of vulnerabilities is not immediate but gradual. Further, IDPs do not constitute a homogenous group, making it complicated to design universal programs for them. At the same time, one must allow them to freely choose the durable solution they prefer for their reintegration in the society. This requires a combining of humanitarian and development approaches (Harild & Christensen, 2010; Harild, 2016).

Energy access, like other basic infrastructures such as ICT services and rural roads, is considered as one of the conditions determining IDPs reintegration in society. However, information is lacking on how energy access can be realized, what needs to be done, and the influence it has on the overall result of programs and policies implemented. There is vast and growing evidence of the development impacts of energy access. Renewable energy and decentralized systems are becoming increasingly relevant in having a positive impact on health, education, and empowerment. Since the needs of IDPs change gradually (Kälin, 2003), we need to explore the possible contribution of renewable decentralized energy systems to achieving sustainable solutions.

On November 29, 2016, Colombia concluded the ratification of the Peace Accords with the FARC-EP.[1] This ended one of the longest internal armed conflicts in Latin America and the world. However, the signing of the agreement was only the first step of an even more complex process toward the total pacification and normalization of the country. Also known as the Havana peace process, this agreement was strongly backed by the international community. The Colombia en Paz Fund was constituted in 2017 to coordinate among the four peace-related funds created by the United Nations, the World Bank, the Inter-American Development Bank, and the European Union (Fondo Europeo para la Paz en Colombia, n.d.). The international community and all kinds of local organizations (public entities, private enterprises, and nongovernmental organizations) also oriented their programs to support the postagreement process.

Colombia's peace programs relate to the sustainable development of the country. It is one of the building blocks of the Peace Accords, with the national development plans of the previous and current governments focusing on the sustainable development of the country, particularly in conflict-prone rural areas. In parallel, the National Determined Contributions submitted by the government toward fulfilling the Paris Climate Agreement stipulations explicitly describe their importance for the peace-building and consolidation processes.

IDPs constitute the largest group of victimized people in Colombia. Achieving durable solutions for them is a top priority for a shift away from the conflict period. The number of victims of the conflict is so high that almost every Colombian is a victim in some way or is in a victimized situation to some extent, as can be seen from the fact that over 7 million IDPs are settled mainly in shantytowns at the peripheries of medium and large urban areas.

The three possible durable solutions are their return, their local integration, or their relocation in another part of the country, although the implementation of these processes would be challenging. Most IDPs prefer to return to their hometowns in the countryside. However, the security conditions are still unstable in some territories, and even in areas where it is safe to return, IDPs would face enormous economic and social challenges. International donors have focused on these rural areas, creating not only a positive momentum but also, in some cases, an "overdose" of cooperation that is not sustainable in the long term.

[1]The Spanish acronym for the Revolutionary Armed Forces of Colombia–People's Army.

At the same time, the younger generations would rather remain and integrate into the host communities. However, the legalization process of their settlements in the peripheries of big cities is complex and could take more than ten years if it happens at all. This would prevent the integration of the IDPs in the formal city development plans and would exclude them from formal access to services. Consequently, these areas become prone to marginalization and higher levels of criminality. Local governments are undertaking efforts to legalize these areas, and numerous innovative solutions are being developed to improve the living conditions in these areas and to integrate the areas into the municipal plans.

This chapter explores how energy projects and programs can have a positive impact on the achievement of durable solutions for IDPs in Colombia. The analysis relies on data collected through a literature review and field study in Colombia. We also conducted semistructured interviews with experts and practitioners in Colombia's energy sector and peace process, and we conducted two discussion focus groups to identify the contradictions between the peace and energy agendas and to find possible interlinkages. All these were conducted in Bogota, the capital of Colombia, and Pasto, the capital of the border province of Nariño, to incorporate perspectives from both urban and conflict-prone areas. All the responses have been anonymized to protect the privacy of the experts consulted.

The next sections summarize the findings by first providing the context of Colombia's peace process in relation to the sustainable development of the country, including the energy access issues, focusing on the challenges for IDPs. This is followed by a description of the international and Colombian frameworks for the support of IDPs. Next, we analyze the contribution of energy projects in the realization of two of the durable solutions (local integration in urban areas and return). Analyzing relocation solutions was not possible, owing to a lack of sufficient data.

4.2 Colombia's peace process, internal displacement, and sustainable development

Colombia's internal conflict has its roots in the context of the Cold War, U.S. interventions in Latin America (Agudelo Blandón, 2015), and the emergence of several armed groups. The groups have been fighting with the government and, between them, have caused millions of civil victims in the process. The study of the armed conflict has been one of the drivers of social sciences in

Colombia. The conflict has been linked with the state-building process of Colombia (Gonzalez Gonzalez, 2014). Lack of free spaces for political participation, socioeconomic exclusion, and other factors converged in the emergence of the guerrilla warfare in Colombia (Pizarro, 1987).

The most prominent guerrilla group, the FARC-EP, was formed in 1964 as a Marxist self-defense group in the countryside (Cosoy, 2016), but it has not been the only active group in the country. The National Liberation Army (ELN), which was formed in 1965, was influenced by the Cuban revolution and the theory of liberalization. In 1966 the Popular Liberation Army was created with links to the Communist Party. A fourth major guerrilla group, the 19th of April Movement or M-19, was created in 1973 with the objective of democratization after the electoral fraud of 1970. Likewise, the paramilitaries and right-wing groups emerged and multiplied in the 1980s. These are "armed actors who, through the use of violence, pursue social, economic and political purposes, supporting and allowing the reproduction of the Colombian social space" (Koessl 2015, p. 21). In the 1990s the AUC was formed through the integration of several paramilitary groups. Table 4.1 shows the armed groups in the Colombian internal conflict.

The drug cartels, which had begun operating in the 1970s, emerged as a major force in the 1990s when Colombia became

Table 4.1 Armed groups in Colombia.

Name of the armed group	Type	Current situation
Fuerzas Armadas Revolucionarias FARC	Guerrilla	Peace agreement in 2016
Ejército de Liberación Nacional (ELN)	Guerrilla	Operational
Ejército Popular de Liberación (EPL)	Guerrilla	Operational
Movimiento 19 de Abril (M-19)	Guerrilla	Peace agreement and demobilization in 1990
Autodefensas Unidas de Colombia (AUC)	Paramilitary	Demobilized in 2003–2006
Grupos armados organizados (Bacrim)	Paramilitary	Not all criminal organizations that operate in Colombia have been identified.
Ejercito Nacional de Colombia	State armed forces	

Source: Based on Cosoy, N. (2016, August 24). ¿Por qué empezó y qué pasó en la guerra de más de 50 años que desangró a Colombia? BBC News. *https://www.bbc.com/mundo/noticias-america-latina-37181413.*

the largest producer of coca leaves in the world. Drug cartels, those in the Medellin and Cali, became prominent criminal groups with links to the paramilitary and guerrilla groups. All of these triggered a strong response from the government, which mobilized the state armed forces. The United States provided vast military support under the so-called Plan Colombia to tackle drug trafficking and armed conflict and to foster economic and social development (Guevara, 2015). However, this support only intensified the conflict (Stokes, 2005).

Internal armed conflict has led to a social catastrophe in Colombia. The *Basta Ya!* report of the Historical Memory Center reveals that between 1958 and 2012, the conflict caused the death of 218,094 people (GMH, 2013). Among these fatalities, 40,787 were combatants (equivalent to 19%), and 177,307 were civilians (81%). The number of missing people between 1981 and 2010 was 25,000, the number of abducted was 27,023, and the number of murders increased to 150,000. The paramilitaries were responsible for 38.4% of the deceased, the guerrilla for 16.8%, and Public Force for 10.1%.

With widespread forced displacement, information published by the Unit for Victims reveals that by December 31, 2014, the Unified Victims Registry (RUV) had registered a historical number of 6,459,501 forced displaced victims. According to the RUV, 1114 municipalities in the 77 regions of Colombian territory have registered cases of forced displacement. This number represents 99% of the national geography; the conflict had displaced at least one person per municipality in Colombia. Even after the peace agreement was signed, displacement continued at alarming levels, owing to the lack of power in the areas that had previously been controlled by the FARC (Asmann, 2019). Table 4.2 and Fig. 4.1 present the evolution of internal displacement.

The conflict-induced forced displacement had a nationwide impact. Rural areas on the Pacific coast and areas controlled by the FARC became the areas to which people were expelled. The displacement created shantytowns in large urban areas. Bogota and Antioquia (Medellin and Cartagena) attracted the largest number of displaced people (see Fig. 4.2). This influenced the urban growth of the cities (Torres Tovar, 2012). In the new scenario the victims of the conflict, millions of people who had traditionally been invisible and marginalized, began to be recognized; they obtained legal rights and guarantees and established reparation mechanisms. Furthermore, even when these neighborhoods become legalized, their vulnerability does not end. Many are still

Table 4.2 Periods of armed conflict in Colombia and forced displacement (Centro Nacional de Memoria Historica, 2015).

Period	Armed conflict	Forced displacement
1948–1958	Violent and bipartisan war between liberals and conservatives due to the death of Jorge Eliécer Gaitán (political leader and presidential candidate).	Violent murders, many of them indiscriminate massacres that left at least 200,000 people dead. Between two and three million people were forced to move.
1958–1974	The National Front: Alliance between the liberal and conservative political parties for alternating power. Inequalities in the countryside led to the formation of the guerrillas (FARC and ELN). President Guillermo León Valencia created civilian self-defense groups to counter the guerrillas.	"The profound effects on the displaced people during the time of violence, the accumulation of land by landowners, the failure of the agrarian reform and the brutal state response during the National Front provided a functional scenario for the insurgent struggle and the proliferation of armed actors."
1974–1979	Irruption of drug trafficking.	Use of land for illicit crops promoted waves of migration of "emerging classes of dubious origin" and of a peasantry without economic options.
1980–1989	Armed conflict escalated with the use of violence legitimized by all the actors.	Exodus of the population as a way of conflict resolution, making war more intense and dirtier.
1989–1996	Demobilization of five guerrilla groups, including M-19 and EPL, and the creation of the Political Constitution of 1991. Remaining guerrillas and paramilitaries became more powerful.	Displacement continued to increase and spread to more regions, particularly Urabá and the Caribbean.
1997–2004	Grouping of different paramilitary groups under the umbrella of the United Self-Defense Groups of Colombia (AUC) and expansion across the country. The violence of the guerrillas (especially the FARC) against the civilian population intensified.	Disproportionate increase in forced displacements. In these seven years, 3,087,173 people were forced to flee their homes.
2005–2014	Demobilization of the AUC between 2005 and 2006. Later, several of its structures, known as Bacrim, were rearmed. Guerrilla actions continued against the increased presence of security forces in territories that were under their control.	Persistence of displacement scenarios during the quest for peace.

Source: Centro Nacional de Memoria Historica. (2015). Una nación desplazada: Informe nacional del desplazamiento forzado en Colombia. http://www.centrodememoriahistorica.gov.co/descargas/informes2015/nacion-desplazada/una-nacion-desplazada.pdf.

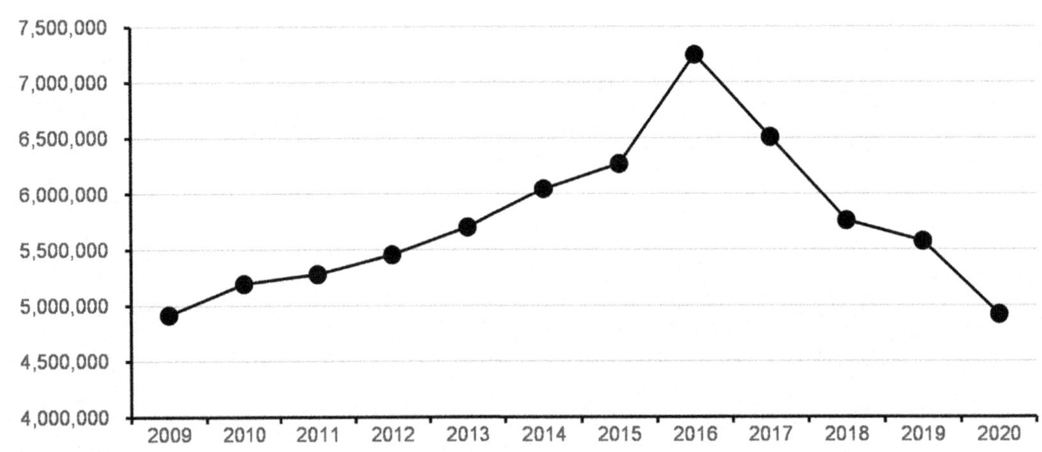

Figure 4.1 Internally displaced persons in Colombia due to conflict and violence. Internal Displacement Monitoring Centre (IDMC).

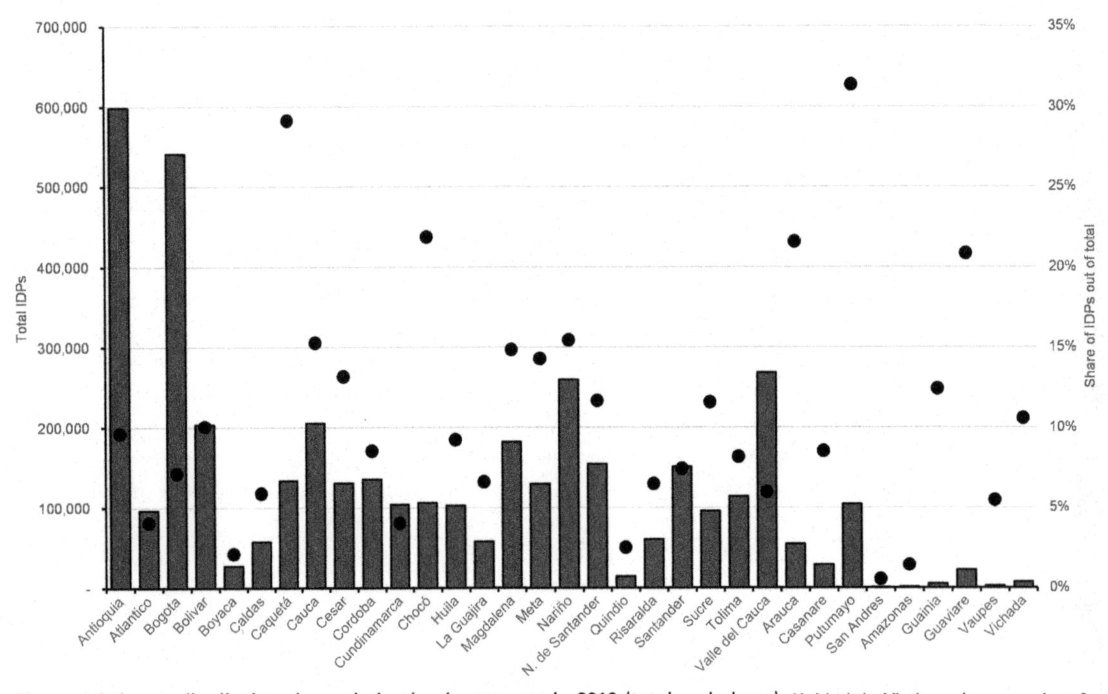

Figure 4.2 Internally displaced population by department in 2012 (total and share). Unidad de Victimas (accessed on 6 August 2021 from https://cifras.unidadvictimas.gov.co/Home/Desplazamiento?vvg = 1).

marginalized and not well integrated into the city; they live in the "peripheries," not just physically but also economically and politically (TECHO, 2015).

4.3 Realizing durable solutions for internally displaced persons in Colombia

4.3.1 Conditions for durable solutions and Colombia's approach

A durable solution for IDPs is achieved when they "no longer have any specific assistance and protection needs that are linked to their displacement and can enjoy their human rights without discrimination on account of their displacement" (IASC, 2010). IDPs are a special type of forced migrants in the sense that they do not cross international borders; hence, in principle, they are entitled to the same rights and freedom of movement as any other citizen. However, this would increase the invisibility of their difficulties. Furthermore, they are not protected by any international convention or humanitarian law (Kalin, 2005).

The Inter-Agency Standing Committee and the United Nations Secretary set the basis for international attention (IASC, 2010; UNSG, 2011), according to which there are eight criteria for measuring the level of achievement of a durable solution (see Table 4.3). More recently, the World Bank report supporting development approaches (World Bank, 2017) and the International Organization for Migration (IOM, 2017) have also determined the criteria. All this has been fostered by a continual increase in the number of forced displaced due to conflict, development projects, and natural

Table 4.3 Criteria to determine the extent to which a durable solution has been achieved.

#1	Long-term safety and security
#2	Enjoyment of an adequate standard of living without discrimination
#3	Access to livelihood and employment
#4	Effective and accessible mechanisms to restore housing, land, and property
#5	Access to personal and other documentation without discrimination
#6	Family reunification
#7	Participation in public affairs without discrimination
#8	Access to effective remedies and justice

Source: International Framework (IASC. (2010). IASC framework on durable solutions for interllay displaced persons. https://www.unhcr.org/50f94cd49.pdf).

disasters. Recognizing the importance of durable solutions as a right of the vulnerable population and the need for the conflict resolution, Koser (2007) states:

> *Resolving displacement is inextricably linked with achieving peace, especially where the scale of displacement is significant. Helping displaced populations to return and reintegrate can simultaneously address the root causes of a conflict and help prevent further displacement. Among the displaced, IDPs often have needs that are different both from refugees and other war-affected civilian populations, and thus they require special attention in peace processes.*

However, realizing any of these durable solutions is a complex process (see Table 4.3), particularly reintegration (Graviano, Götzelmann, Nozarian, & Wadud, 2017; Samuel Hall, NRC, & IDMC, 2018).

Over the years, Colombia has developed an institutional architecture for supporting the victims of the conflict with the support of civil society organizations. The 2011 Victims and Land Restitution Law (Ley 1448 de 2011, 2011) is a milestone in this process (Ley 1448 de 2011, 2011). The norm includes the restitution of millions of acres to displaced people and financial reward for victims of violations of human rights. For its execution a series of decrees were issued, and in January 2012 the Unit for Integral Attention and Reparation for Victims was created, which works in coordination with the National System of Comprehensive Care and Reparation for Victims integrated by public entities at the national and territorial level and other public or private organizations that are a part of the efforts of the Unit to achieve integral reparation for victims.

From a transformative perspective, comprehensive reparation requires transversal support that cuts across economic, social, and cultural change spheres. "The reparation must try to respond to the different dimensions of the damage suffered by the victims, which requires a multidimensional approach, which is not limited to the delivery of monetary compensation" (Portilla Benavides & Correa, 2015). Beyond compensation the integral reparation of victims refers to the "right to be repaired in an adequate, differentiated, transformative, and effective manner. Reparation includes the measures of restitution, compensation, rehabilitation, satisfaction and guarantees of nonrepetition, in its individual, collective, material, moral and symbolic dimensions" (Article 25, Colombian Congress, 2011).

Another significant step for the recognition of the victims and their rights was the signing of the General Agreement for the Termination of the Conflict and the Construction of a Stable and Lasting Peace in Colombia between the government and the FARC guerrillas (August 24, 2016). The main aim of this agreement was to grant victims' rights; to ensure accountability for what happened; to guarantee legal security to those who participated in it; and to guarantee coexistence, reconciliation, and nonrepetition as essential elements in the transition to peace.

Within the scope of the Integral System of Truth, Justice, Reparation, and Non-Repetition, five mechanisms were defined: the Commission for the Clarification of Truth, Coexistence and Non-Repetition, wherein there is no criminal imputation of those who appear before her; the Special Unit to search for missing persons in the context of the armed conflict; the Special Jurisdiction for Peace, which will exercise judicial functions in the face of serious infractions of IHL or violations of Human Rights; Measures of integral reparation for the construction of peace, including the rights to restitution, compensation, rehabilitation, satisfaction, and nonrepetition; and Guarantees of No Repetition, which is the outcome of implementing previous mechanisms, and is in compliance with the other points of the peace process agenda (Center for Studies for Peace - CesPaz, 2017).

4.3.2 Piloting durable solutions: the transitional solutions initiative by United Nations High Commissioner for Refugee and United Nations Development Programme

The United Nations High Commissioner for Refugees (UNHCR) and the United Nations Development Programme (UNDP) launched the Transitional Solutions Initiative (TSI) between 2012 and 2016 (Gottwald, 2016). This was an initial attempt to combine humanitarian and development approaches to realizing durable solutions for the displaced population. This was aligned with the approval of the Law of Victims and Land Restitution by the national government, which began the institutional development to support the victims of the conflict. In contrast to the official approach, the TSI was designed to incorporate active participation by the community, both the displaced individuals and the host community (Econometria, 2016).

The TSI covered 17 communities with a variety of affected communities and pilot solutions across the three possible durable solutions (return, relocation, and local integration) (see Table 4.4). Piloting context-specific solutions for each of the communities, the TSI brought better understanding to the needs and actions for

Table 4.4 Overview of UNHCR-UNDP transitional solutions initiative (TSI) projects.

Community	Town	Department	Type	Budget (USD)	Main executer	Houses
Loma Central Alta Montaña	Carmen de Bolivar	Bolivar	Return	212,752	UNDP	590
Casacará	Agustin Codazzi	Cesar	Return	212,752	UNDP	2,000
Tanguí	Medio Atrato	Chocó	Return	248,017	UNHCR	272
El Arrayan + La Argentina	Nariño	Antioquía	Relocation	160,888	UNHCR-UNDP	21
Las Delicias and El Rodeo	Puerto Lopez	Meta	Relocation	894,361	UNHCR-UNDP	128
Chami Puru (Embera)	Florencia	Caquetá	Relocation	757,337	UNHCR	51
Nasa Páez community	Florencia	Caquetá	Relocation	395,109	UNHCR	14
Resguardo Edén (Awa)	Ricaurte	Nariño	Relocation	743,590	UNHCR	132
Altos de la Florida	Soacha	Cundinamarca	Local integration	1,531,686	UNHCR-UNDP	625
13 de Mayo	Villavicencio	Meta	Local integration	1,349,433	UNHCR-UNDP	1,022
Vereda Granizal	Bello	Antioquía	Local integration	1,543,882	UNHCR-UNDP	3,600
Nueva Esperanza	Mocoa	Putumayo	Local integration	783,788	UNHCR	228
Manuela Beltran	Cucutá	N. de Santander	Local integration	720,052	UNHCR	642
Las Delicias	Cucutá	N. de Santander	Local integration			613
La Gloria (El Puerto)	Florencia	Caquetá	Local integration	531,669	UNHCR	234
Villa España	Quibdó	Chocó	Local integration	295,366	UNHCR	94
Familias en Acción	Tumaco	Nariño	Local integration	405,320	UNHCR-UNDP	196

Source: Econometria. (2016). External Evaluation of the Joint UNHCR – UNDP Program "Transitional Solutions Initiative - TSI" (Issue October). http://tsicolombia.org/sites/acnur/files/descargas/executive_summary_tsi_evaluation_colombia.pdf.

the development of durable solutions in Colombia. Another TSI program was implemented by the UNHCR and UNDP in Eastern Sudan.[2] The Korea International Cooperation Agency (KOICA) and UNHCR initiated a follow-up program building on the TSI for

[2]http://www.europe.undp.org/content/geneva/en/home/library/crisis_prevention_and_recovery/undp-unhcr-transitional-solutions-initiative--tsi--joint-progr.html.

urban solutions in Colombia (Unidad para la Atención y la Reparación Integral a las Víctimas, 2019).

The initiative provided several lessons, despite projects planned from their inception as pilot projects with unique characteristics. A key challenge was the complexity of producing replicable solutions, since flexibility was crucial in adapting the solutions to each community. However, some general findings were to be considered for any initiative to support the realization of durable solutions for IDPs. These include community empowerment and ownership, nondiscriminatory interventions, and the active incorporation of women and youth (see Econometria, 2016 p. 82 for a complete and detailed description).

4.4 Energy access programs to spur durable solutions for internally displaced people in Colombia

Energy projects in Colombia have been closer to conflict than to peace. In rural areas, extractive mining, hydropower projects, and oil pipelines have been associated with forced displacement. The development of large-scale dams has complex social and environmental impacts, including the displacement of populations (Siciliano, Urban, Tan-Mullins, & Mohan, 2018). The 2456-MW Hidroituango project (or Ituango dam project) is expected to be the largest in the country, across the Cauca River. The dam will contribute to increasing low carbon generation to the mix and will mobilize economic programs to foster development and minimize local impacts. However, there have been several complications, including accidents during construction (El Espectador, 2018) and public criticism resulting from the social and environmental impacts (Torres, Caballero, & Awad, 2016). Oil pipelines have also been the target of criminal groups (Semana, 2002). While the government has been able to dramatically reduce this problem, attacks on small, isolated communities are difficult and costly to prevent. In urban areas, energy access in peripheral areas continues to be an unresolved issue. The legalization process of informally settled areas is complex, and it can take decades for the local municipality to finish all the procedures. In many cases, these projects might not be feasible because of risks of landslide or occupation of protected areas. The crisis of Electricaribe on the Caribbean coast is one of the most salient examples. In fact, a feeling of inclusion in the modern city through highly visible actions as cable cars is necessary (Brand & Dávila, 2011).

However, renewable energy projects and decentralized energy planning can shift this paradigm. A growing literature on the impact of energy access on individuals and communities goes beyond electricity and/or cooking access per se and examines the role of decentralized solutions (Baldwin, Brass, Carley, & Maclean, 2015; Brass, Carley, MacLean, & Baldwin, 2012; Cabraal, Barnes, & Agarwal, 2005). Development academia and practitioners have often pointed out the educational and health benefits. The effects on women's empowerment are also of great relevance (Listo, 2018; Mahat, 2011; Samad & Zhang, 2019). Hence the need for different approaches and business models has been widely recognized (Chaurey, Krithika, Palit, Rakesh, & Sovacool, 2012; Sovacool, 2013).

Universal access to modern energy services is one of the goals of the United Nations Agenda 2030 for Sustainable Development (SDG 7). More recently, the level of energy access and the process for it have been considered important. The World Bank energy program led to the creation of the Multi-tier Framework for Energy Access to "monitor and evaluate energy access by following a multidimensional approach" (Bhatia & Angelou, 2015; Bhattacharyya, 2012; Rysankova, Portale, & Carletto, 2016). All this has led to an increasing interest in incorporating energy democracy, justice, and sovereignty into the assessment of energy systems and their overall sustainability (Burke & Stephens, 2017; Sovacool & Dworkin, 2015).

In general, energy is considered a required input to ensure a good quality of life and economic opportunities as part of durable solutions for IDPs. The link between energy and peace building is subtle. Considering the transition of energy systems worldwide, it is important to focus on how better access to energy can address the vulnerabilities of displaced populations and can be considered a potential form of reparations (Souter, 2014). Otherwise, IDPs may remain excluded from the process. Nevertheless, it has not been until recently that energy research has focused on the promotion of peace (Kenner, 2017; Marijnen & Schouten, 2019) and the positive impact on displaced populations. Extensive literature exists on energy projects inducing displacement. Chatham House is a pioneer in the Moving Energy Initiative highlighting energy access for refugees in camps (Lehne, Blyth, Lahn, Bazilian, & Grafham, 2016). Energy peace partners advocate the use of Peace Renewable Energy Credits (PREC) (Mozersky & Kammen, 2018). GIZ held a conference jointly with UNEP and UNHCR on "Energy for Displaced People."[3]

[3]http://sdg.iisd.org/news/giz-and-partners-host-conference-on-sustainable-energy-for-displaced-people/.

In Colombia the 2016 Peace Accords includes sustainable development as one of its pillars. The agreement is based on six points: an integral rural reform, political participation, a ceasefire and the end of hostilities, illicit drugs, victims, and mechanisms for its implementation and verification. Sustainable development is an intrinsic part of all these pillars. The Development Programs with Territorial Focus (Programas de Desarrollo con Enfoque Territorial, or PDET) are an instrument for the implementation of peace-building policies after the agreement. Sustainability principles are essential and not only address environmental concerns, but also become an instrument for the involvement of all the affected communities (including the different ethnicities that are present throughout the country).

However, the peace agreement may also negatively affect the sustainable development of the country. Before their demobilization, FARC-EP controlled vast areas of high environmental value. In their absence, other armed groups fight for the control of such areas. This can lead to an increase in drug production and attacks over communal leaders raising their voice and may thus jeopardize the consolidation of peace. There is also the risk from the development of new extraction and energy-related projects in areas that were previously inaccessible because of the conflict.

Moreover, the populations that have lower levels of energy access are the most vulnerable to climate change (Baker, 2012). The impact will be widespread; 100% of the municipalities of the country will be affected, with 25% severely impacted by 2040–2050 (IDEAM, 2017). The energy mix is also highly dependent on hydropower generation, making it therefore sensitive to changes in El Niño. Climate change could lead to a reduction of about 15% of the hydropower generation, which, together with the difficulties for new projects and increasing energy demand, can lead to an increase in greenhouse gases emissions if the alternatives are more carbon-intensive forms of power generation (Arango-Aramburo et al., 2019). Rural areas and shantytowns in the peripheries of cities that host the population with lower economic resources are the ones that will be more affected. Climate change is already being perceived by rural communities, particularly those near former glacial areas, but the information provided for adaptation is not sufficient, and responses are mostly spontaneous (Barrucand, Giraldo Vieira, & Canziani, 2017). The country's climate commitment in the National Intended Contributions has already emphasized the importance of the connection between climate change and peace-building efforts.

An approach that is focused on sustainable development and clean energy can also be a channel to mobilize international funding sources. U.S. international aid is one example of such a shift. While in the early 1990s, Plan Colombia brought vast military aid to Colombia, this trend has shifted to providing major support to economic areas with the Peace Colombia Plan. Development partners have also created special funds to support the peace efforts. Other international donors too have mobilized funds to support the peace-building efforts. The U.N. Migration Multi-partner Trust Fund, the EU multidonor funds, the World Bank, and the Inter-American Development Bank have all committed to supporting peace building. The United States has also shifted its cooperation toward economic programs, reducing military aid (see Fig. 4.3).

The following sections explore the opportunities for energy access projects to contribute to the realization of durable solutions for IDPs based on a survey conducted in Colombia, including semistructured interviews with experts from government, international organizations, bilateral donors, academia, private sector, and civil society actors. In addition, two focus group discussions were conducted in Bogota and Pasto in which participants shared their views on the linkages between energy access, renewable energy, and peace consolidation. The next two sections outline the main findings, looking at solutions for the integration of displaced population in rural areas (returnees) and urban areas (local integration). Throughout the survey, very limited cases of relocation were found. This topic is not included in this chapter, although it is one that should be further investigated.

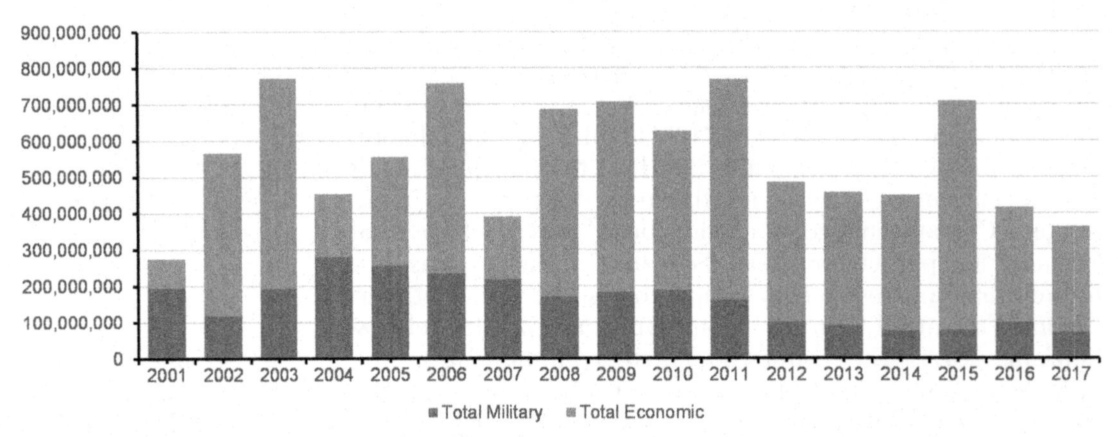

Figure 4.3 U.S. disbursed foreign aid to Colombia. U.S. Foreign Aid Explorer: https://explorer.usaid.gov/cd.

4.4.1 Promotion of rural electrification in Colombia and the return to rural areas

The government of Colombia has been supporting return as a durable solution under the principles of willfulness and security. The government programs include support for land acquisition and initial settlement, including access to public services. However, younger generations are reluctant to return to rural areas where job opportunities are scarcer than in the big cities. Securing economic and development opportunities for the returning populations is a basic condition for the programs to succeed.

Access to reliable and adequate energy services is essential for any economic and day-to-day activity. In general, the level of access is lower in rural areas, precariously low in remote areas such as the Choco. Nearly 2 million people in rural Colombia lack access to modern energy services, about 4% of the total population. This population is concentrated in the Non-Interconnected Areas (Zonas No Interconectadas, or ZNI), hovering around 52% of the territory (Superservicios, 2017). These are also the areas with the highest levels of unsatisfied basic needs (see Fig. 4.4). The government has been actively supporting the electrification of this population, mostly with diesel generators. Military clashes, areas controlled by armed

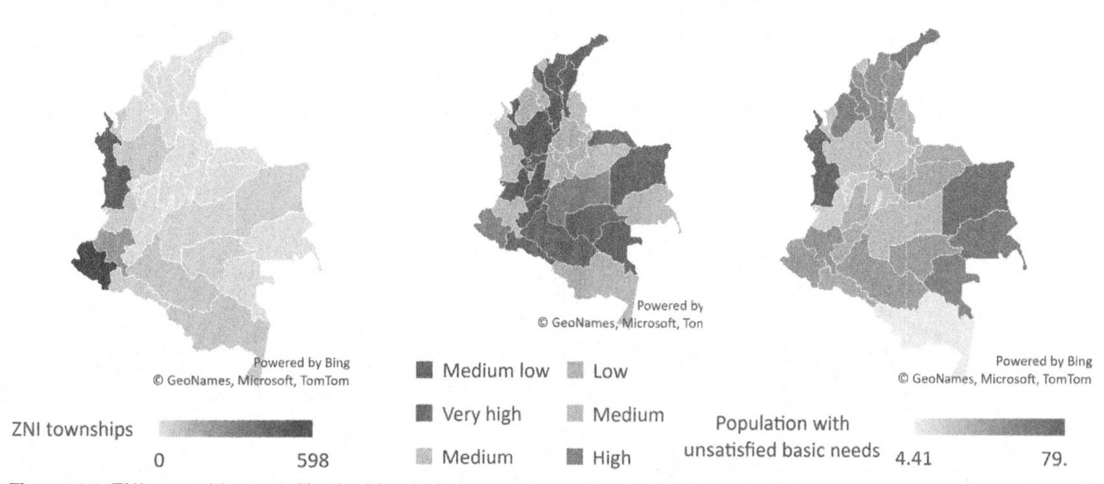

Figure 4.4 ZNI townships, conflict incidence index, and unsatisfied basic needs per department. Indice de Incidencia del Conflicto Armado (IICA): Index with consideration to armed clashes, killings, kidnappings, landmines, forced displacement, and coca cultivation. Adapted by the authors from different sources. ZNI: MinTIC; IICA: DNP; Unsatisfied basic needs: DANE.

groups, and geographical barriers have prevented the expansion of the national grid to a large part of rural Colombia.

Furthermore, the conflict has also severely damaged the energy infrastructures in rural areas. For a long time, transmission grids were the target for guerrilla groups. Attacking pipelines and oil theft by right-wing paramilitary and guerrilla groups became a matter of concern for the government and a matter of revenue for the armed groups (McDermott, 2004). Although the government has made all efforts to tackle this issue, attacks persist (Reuters, 2018). Increases in oil prices in international markets have been correlated with increases in the number of paramilitary attacks (Dube & Vargas, 2013).

Areas that are not connected to the national grid (ZNI) rely mostly on diesel generators running a few hours per day (see Fig. 4.5). Increasing energy supply in rural areas is a prerequisite to improving the quality of life and economic prosperity of the rural population. The government of Colombia is channeling financial resources for the provision of electricity supply to the ZNI areas. Residents with lower income levels are entitled to a subsidy for a minimum level of service, depending on the size of the town (see Table 4.5).

The Peace Accords also includes provisions in this direction. PDETs are the main government strategy for this purpose. These are targeted to the 170 municipalities that are more affected by violence, crops for illicit use (94.5% of the coca plantations were located in PDET municipalities by 2016 (DNP, 2016)), and the lack of presence of the national government (Forero, 2019). These plans were crafted with local communities to identify their needs.

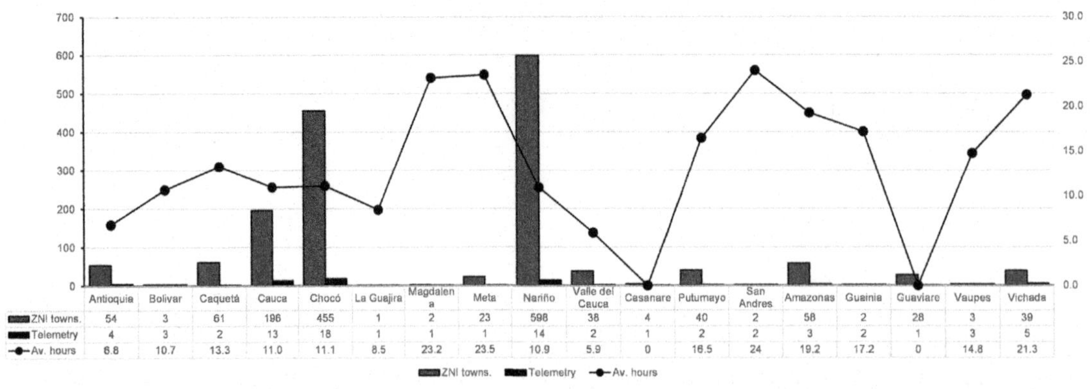

	Antioquia	Bolivar	Caquetá	Cauca	Chocó	La Guajira	Magdalena	Meta	Nariño	Valle del Cauca	Casanare	Putumayo	San Andres	Amazonas	Guainia	Guaviare	Vaupes	Vichada
ZNI towns.	54	3	61	196	455	1	2	23	598	38	4	40	2	58	2	28	3	39
Telemetry	4	3	2	13	18	1	1	1	14	2	1	2	2	3	2	1	3	5
Av. hours	6.8	10.7	13.3	11.0	11.1	8.5	23.2	23.5	10.9	5.9	0	16.5	24	19.2	17.2	0	14.8	21.3

Figure 4.5 ZNI townships with installed telemetry and average hours of electricity service (March 2017). MinTIC https://www.datos.gov.co/Minas-y-Energ-a/Estado-de-la-prestaci-n-del-servicio-de-energ-a-en/y2yi-4r6y/data.

Table 4.5 Government Estrato subsidy for electricity consumption in ZNI.

Number of consumers	0–50	4 hours/day or 33.6 kWh/month
	51–150	5 hours/day or 46 kWh/month
	151–300	8 hours/day or 77 kWh/month
	>300	Average of Estrato 1 consumption in the same department
San Andres, Providencia, and Santa Catalina Archipelago		187 kWh/ month
Total 2016 subsidy		255,564,578,753 COP (79,944,433 USD)

Estrato is Colombia's mechanism for subsidy allocation. The scale is from 1 to 6, with 1 being the lowest and 6 the highest "expected" income groups; 1 to 3 are entitled to benefits, 4 are expected to pay for their use, and 5 and 6 overpay so that 1 to 3 can be offered the subsidy. The classification is not based on actual income or socioeconomic conditions, but through an assessment of the quality of their housing.
Source: Superservicios. (2017). Diagnostico de la calidad del servicio de energia electrica en Colombia *(Issue May).*

The Peace Accords include a commitment from the government for the National Rural Electrification Plan.

The Program for Sustainable Rural Energization (Programa de Energizacion Rural Sostenible in Spanish, or PERS) aims to develop rural electrification plans in the ZNI with a strong focus on the development of productive uses. The program promoted by the Instituto de Planificación y Promoción de Soluciones Energéticas (IPSE)[4] and Unidad de Planeación Minero Energética (UPME)[5] was launched with support from the U.S. Agency for International Development (USAID) Colombia Clean Energy Program. It promotes rural electrification in ZNI through the engagement of local communities, governments, and universities. The idea behind this was to promote more bottom-up approaches to the design of better solutions for the specific context of each department. The involvement of regional universities and governments also supports the increase in capacities outside national entities (UPME and IPSE). Surveys were conducted for each department to assess both the existing energy resources and the needs of the population (see Table 4.6). The data that were gathered would help in the development of energy systems to foster economic activities. Indeed, this is an essential component to secure the sustainability of systems in the medium and long terms. The program is moving toward the assessment of viable business models to involve communities, private actors, and local governments.

The involvement of local universities is also instrumental in increasing the adequacy of the solutions as well as fostering

[4]In charge of noninterconnected areas.
[5]In charge of the national interconnected system.

Table 4.6 PERS projects by department.

PERS	University	Survey	IPSE Funding (COP million)	Link
Guajira	SENA			http://persguajira.corpoguajira.gov.co/
Tolima	U. Tolima			http://perstolima.ut.edu.co/
Nariño	Universidad de Nariño	3004	40	http://sipersn.udenar.edu.co:90/sipersn/
Chocó	Universidad Tecnológica del Chocó	2960	140	http://www.perschoco.com/
Cundinamarca	Universidad Distrital	1375	240	http://egresado.udistrital.edu.co/
Orinoquía	Universidad de los Llanos	3986	3,500	http://observatorio.unillanos.edu.co/pers/
Norte de Santander	Universidad Francisco de Paula Santander	2228	1,300	http://persnds.ufps.edu.co/pers_app/public/
Cesar	Gobernación		500	http://cesar.gov.co/d/index.php/es/menpre/menprenoti/334-artbp-0823-2016
Putumayo	Universidad de Nariño	1411	650	http://persputumayo.udenar.edu.co

Source: IPSE http://www.ipse.gov.co/pages/ipse/Informe_PERS_Direcci%C3%B3n1.pdf.

further development of other sustainable initiatives. For example, the University of Nariño, which is in charge of the PERS of the department of the same name, launched new energy programs, such the use of electric bicycles for students living far from the main campus, under the program Campus Verde.[6]

However, IDPs who are considering return find multiple challenges beyond economic development. Security and land issues are the most immediate needs. Provision of public services is not only a public responsibility but also a mechanism for reinforcing the state's presence over the territory and contributes to peace building. The demobilization of FARC-EP armed forces left a vacuum of power in many areas. The government has found it difficult to control these areas. As a result, communities find themselves without protection and in the context of continuing conflict.

The dependence on the supply of diesel (see Fig. 4.6) for generating electricity for a few hours daily diminishes the self-reliance of the returnee community. The generosity of the government subsidy can create vested interests and foster a dependency relation. Furthermore, drug cartels use oil and gasoline intensively in the coca production process. As a ballpark figure, 1680 L of

[6]See http://facultades.udenar.edu.co/proyecto-campus-verde-udenar/ (in Spanish).

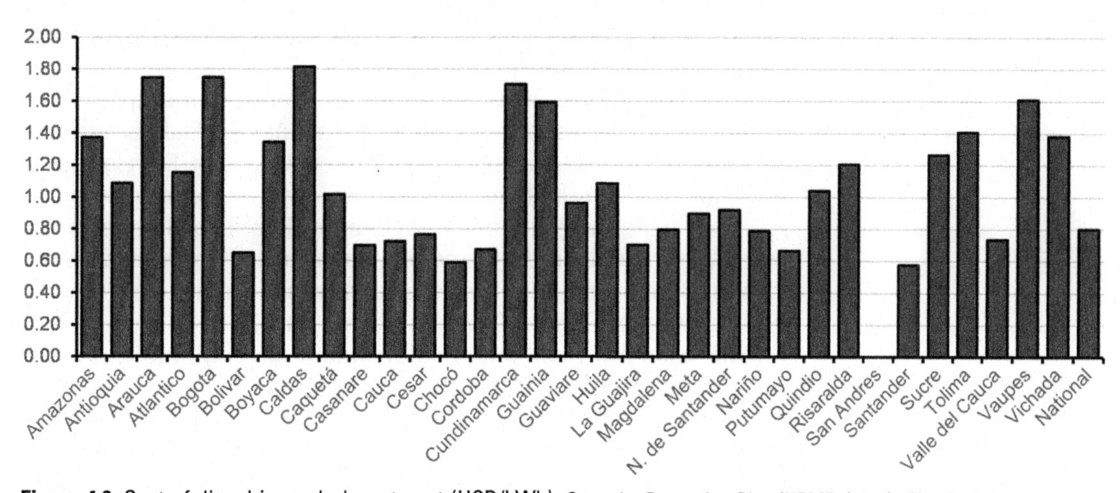

Figure 4.6 Cost of diesel in each department (USD/kWh). Capacity Expansion Plan (UPME. (2014). *Plan Indicativo de Expansión de Cobertura 2013–2017*).

gasoline are needed to produce 110 kg of coca (Hernandez, 2019). As was mentioned earlier, the diesel supplies to isolated communities are potential targets of attack. This reduces the available resource for the limited hours of energy supply. Furthermore, the sense of vulnerability hinders the stabilization of the population.

Returnees face complex psychological and emotional stress when going back to the place where they were victimized. Displacement has already caused severe damage to the social fabric of the communities. In cases in which IDPs return to their own rural areas, forced displacements divided the society into the ones who stayed and persisted and the ones who abandoned and returned, thus generating resentment, anger, and conflict within the community. Forced displacements fragment cultural identity and promote segregation and discrimination toward the IDPs who return to their own community.

A Colombian expert in the peace process pointed out that "a peace approach requires the transformation of these communities in the democratic processes, developing capacities for defending their own territorial peace and being able to start a dialog with the State and their own threats. [Therefore] people have to re-learn to govern themselves and participate in the Government."

Research has shown that focusing not only on the provision of energy access but also on how it is provided can help to generate or regenerate a governance structure and contribute to the stabilization of communities. The accomplishment of goals involving a full provision of the population's needs reinforces social links, recreates community identity, and leads to the generation of new governance structures wherein people acknowledge their own capacities and are able to negotiate for their future.

A plethora of initiatives are aimed at fostering community ownership and involvement, creating independent governance systems. Development partners, such as the Inter-American Development Bank and the World Bank, are also involved in promoting such schemes with programs such as Somos Pazcifico. Initiatives in other sectors find similar outputs. A communal system for the collection of rainwater and its distribution helped strengthen the local Junta de Accion Comunal (Valdés Correa, 2017).

The establishment and reinforcement of community governance mechanisms is both an input and an outcome of a common good project. The sustainability of an electrification project is highly dependent on the ability of the community to operate and maintain it. A sense of ownership is essential as well as a close connection with tangible benefits to the community. The mini-hydropower plant in Palmor, Magdalena, is a good example of this. IPSE supported the strengthening of the energy supply in the community through the USAID Clean Energy Program. Building on an existing local cooperative and a small project, the program strengthened all the systems and provided the required technical training for the staff. Currently, the project is managed independently by the community.

Another aspect that can enhance security conditions is to foster a change in mindset. The length of the conflict has created the social dynamics of conflict in many areas. Populations have become habituated to a climate of violence. One of the interviewees expressed it thus: "We have learned that we can resolve issues with weapons, with violence. Changing this mindset requires a strong change in our social imaginary." This has hindered the integration of vulnerable populations. Programs for the youth, such as IPSE's Centinelas de la Energia, are promising approaches to building new scenarios for the future. The program targets the younger generations to incorporate them into energy management. By understanding the systems and the merits of efficiency measures, their involvement levels become higher.

Another element for improving security is the demobilization of combatants and their integration into society. Energy projects and clean technologies are attractive options for demobilized combatants to integrate them socially and economically in a society in peace. An interesting case is the development of the Archimedes Screw by former FARC-EP militia in one the camps that had been assigned by the national government as a temporal shelter for them (Calle, 2018).

4.4.2 Electricity services in the peripheries and the local integration of internally displaced people

Many displaced persons have moved to the largest urban areas in Colombia, and this has led to the creation of numerous informal settlements or "barrios de invasion." In 2003, more than a million households were estimated to live in informal settlements (Torres Tovar, 2009). In Bogota, some districts have as high as 21.8%, 21.2%, or 18.5% of informal households (Santa Fe, Ciudad Bolivar, and La Candelaria, respectively) (Secretaría Distrital de Hábitat, 2017). The growth of informal settlements by IDPs has spread to neighboring towns. For example, Soacha (Cundinamarca), across the southern border of Bogota, had received 54,017 IDPs by 2017, with 12,512 households (Secretaria de Planeacion y Ordenamiento Territorial, 2017). The legalization of the settlements is a priority for the local governments, but it is a slow and complex process. In 2018, Bogota legalized 25 neighborhoods with 8300 residents (Sabogal, 2019). Legalization is essential to overcome the vulnerabilities of IDPs (Colombia2020, 2018).

In urban areas, access to electricity services is widespread. Even in nonlegalized neighborhoods, residents manage to find ways to connect to the local distribution networks. This has led to nontechnical losses due to electricity theft in informal settlements. Local utilities are not allowed to expand their services to areas that are not legalized; therefore residents in these areas have very few options. The lack of a formal electric connection leads to higher vulnerability to power shortcuts and overheating of the systems (de Bercegol & Monstadt, 2018). Improved energy services have been found to be linked to higher aspirations (Parikh, Chaturvedi, & George, 2012). Nevertheless, the situation varies from city to city. In Bogota, 81.6% of the residents in informal settlements have access to electricity through a metered connection, while 16.8% are connected irregularly, and the remaining 1.6% have no access at all (TECHO, 2015).

Electricity theft is a serious problem, especially in the Caribbean area. The government estimated the figures of "energetically subnormal neighborhoods" to classify those connecting to the local grids without permission (Martínez Ortiz et al., 2013). In 2014 the estimate was 450,060 consumers, about 4% of the total (UPME, 2014), with a high concentration on the Caribbean coast (see Fig. 4.5). In 2006 the government of Colombia approved the Program for the Normalization of Electric Grids (PRONE) for regularizing electricity service (Law 1117 of 2006). In the period 2010–2012, nearly 65 million USD were dedicated to PRONE (UPME, 2014) (Fig. 4.7).

Even in legalized areas, establishing formal connections might be difficult. IDPs find themselves in many cases without the abilities required for jobs in urban areas. Their incomes are minimal and highly dependent on allowances from the government. Their ability to pay for electricity services is critical. The government of Colombia has set a tariff scheme with cross-subsidization that reduces the burden on low-income consumers. Nevertheless, the monthly expenditure on electricity as a percentage of household income is much higher for poorer residents (see Fig. 4.8).

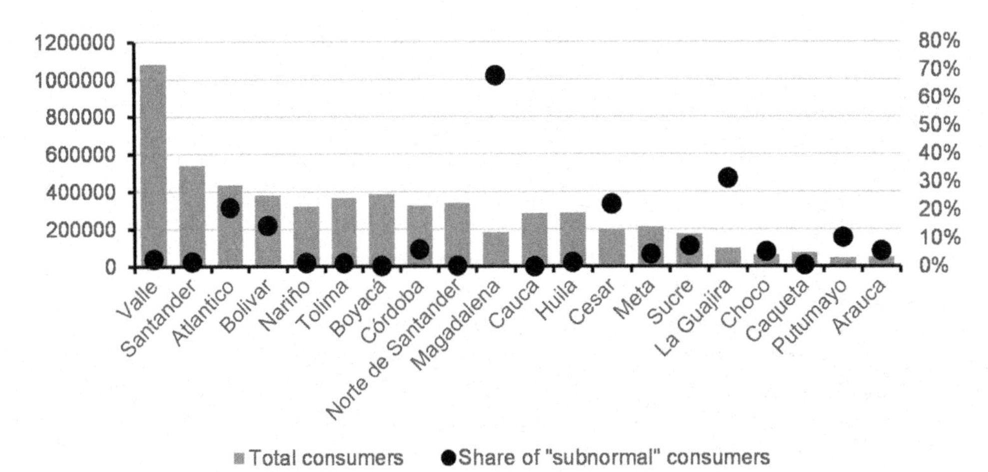

Figure 4.7 Share of "subnormal" consumers. Adapted from UPME. (2014). *Plan Indicativo de Expansión de Cobertura 2013–2017.*

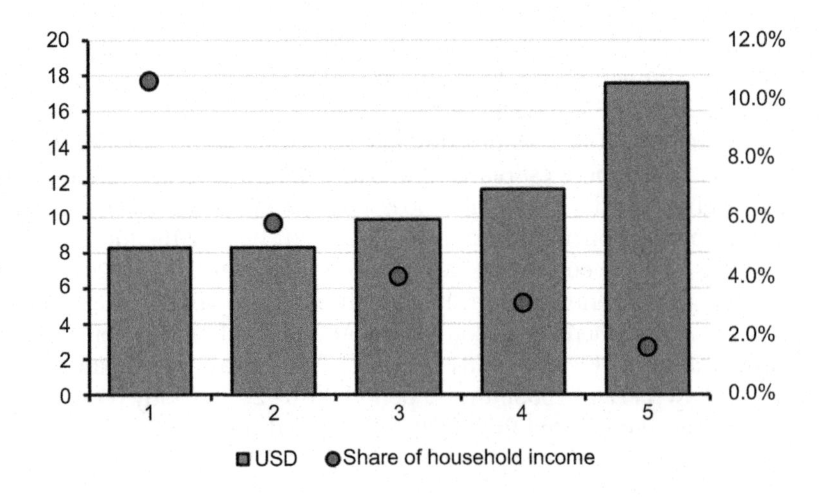

Figure 4.8 Monthly expenditure on electricity as percentage of household income per quintile in Colombia (%) (Komives, Foster, Halpern, & Wodon, 2005). Adapted from Komives, K., Foster, V., Halpern, J., & Wodon, Q. (2005). Water, electricity, and the poor: Who benefits from utility subsidies? http://documents.worldbank.org/curated/en/606521468136796984/Water-electricity-and-the-poor-who-benefits-from-utility-subsidies.

Improved electricity services in the peripheral areas, both informal and legalized, can have an impact on the economic conditions and the quality of life. Electricity is commonly the first formal service that is introduced with the legalization of settlements. Bogota's electricity utility, Codensa, has implemented an innovative mechanism for initial engagement with the population in a neighborhood that is being legalized. Since these areas were previously connected to the service illegally, changing the mindset of consumers to start paying for the service is another complex issue. Community projects such as soccer fields and parks can help in gaining the confidence of the local dwellers. Codensa moved one step further and launched a debit card known as "Easy Credit" (Arbelaez, Garcia, & Sandoval, 2007; Leon, 2016). This was the first opportunity for many people to gain access to financial services. Although it is reported to have low levels of default, the main merit is not the economic returns but the establishment of a relationship between the community members and the local utility.

Shantytowns also suffer from a poor social fabric. The massive displacement can be a cause for discrimination and an inability to overcome the vulnerability conditions. Energy projects, as well as other development projects, can help to reduce the gaps between communities. The nongovernmental organization One Litre of Light has been implementing projects in both urban and rural areas. The first product that it introduced was a bottle of plastic that reflects the light of the sun inside self-constructed houses that lack windows. Originally from the Philippines,[7] this product has become a success in Colombia, and the founder has been awarded several social entrepreneurial awards. Later, they developed a low-cost self-constructed streetlight. The onus of implementing these projects is always on the local inhabitants, who start to work together on a project and gain community ownership over it.

4.5 Summary

In this chapter we examined the role of energy projects, mostly relating to access to electricity, to realize durable solutions for IDPs in the context of the period after peace agreements in Colombia. New alternatives have emerged from the diffusion of decentralized and renewable power generation systems as well as from the development of new business models

[7]https://unfccc.int/climate-action/momentum-for-change/lighthouse-activities/a-litre-of-light-philippines.

for the supply of the service. Durable solutions are almost impossible to achieve in the heat of conflict. However, with the signing of peace agreements, opportunities open up for the displaced population and for the country to seek some form of final settlement (return, local integration, or relocation).

Colombia is currently passing through this phase. The signing and ratification of the Peace Accords in 2016 led the country into a new era. For IDPs this was an opportunity to end their displacement situation and finally settle and integrate, either in the urban areas to which they migrated, back in their home regions, or in a new part of the country. The challenges that lie ahead are not easy to solve.

Our investigation and survey of experts in Colombia clarified several mechanisms in terms of how energy projects can contribute to the realization of durable solutions. The strengthening of IDPs, at both individual and collective levels, is key to reducing their vulnerabilities. This is essential for their social integration and for the reconstruction of the social fabric, thereby improving their quality of life and the economic opportunities that are commonly associated with improved energy access.

IDP settlements, in both urban and return areas, are highly dependent on support from outside. This situation is worsened by the inappropriate implementation of international aids that generate conditions of "assistentialism" (Goulet, 1978), transforming these communities into mere recipients of help without developing their own capacities for resolving their problems. These programs do not tackle the root of the problems that are common in Colombia. Consequently, there is increasing dependence on external support (which might be unrealistic and/or unsustainable in the long term). The dependence model is also followed by the national government for providing access to public services, and the provision of diesel becomes a target for criminal groups that use it in the manufacturing process of illicit drugs.

In rural areas, energy projects represent an opportunity for income generation that will help the population to consolidate in the territory. However, these are separate policy areas in Colombia (agriculture and energy). The involvement of the private sector in the redevelopment can help to bridge the gap, but conflict-prone areas do not attract investors. The government of Colombia initiated a "Work for taxes" ("Obras por Impuestos" in Spanish) program, which incentivizes investment in public works in areas that were affected by the conflict. Development partners such as the World Bank, the Inter-American Development Bank, and bilateral donors are examining this angle. An interesting example is the Spanish Institute for Foreign Trade (ICEX), which launched a program in Colombia to connect Spanish social entrepreneurs with rural communities in the areas in conflict. Although the program does not directly target IDPs, the beneficiaries are likely to have been displaced in the past. ICEX is not an international cooperation agency, as is the Agencia Española de Cooperación Internacional para el Desarrollo.

Local integration in urban areas goes beyond the formal legalization of settlements. The social, economic, and political integration of displaced and host communities is essential for local integration in its truest sense. In the absence of these, the vulnerability of the IDPs will continue. Electricity illustrates this process. It is the first utility service that reaches settlements. Although illegal connections are common initially, this can be attributed to the impossible task of providing service in areas that are not legalized. Nonetheless, as the settlements begin their legalization process, electricity services become the first interaction between the population and the formal mainstream public services provision. The expansion of the service to these areas must be done with consideration of the particularities of the residents. Low income levels are common; therefore affordability must be considered along with other vulnerabilities. The reinforcement of the local social fabric, an appropriate trust-building process between the utility service and the residents, and improvement in the overall image of the settlements are some of the goals that the program for the expansion of electricity services can achieve. Furthermore, this could help in improving the image of the neighborhoods, which are commonly not frequented by other citizens. Infrastructure projects such as the outdoor escalators in Medellin and cable cars in Medellin and Bogota are already enhancing the quality of life as well as opening these areas to local tourists (Bocarejo et al., 2014). The impact of these initiatives in the overall physical and social integration into the city must be considered.

Energy access projects can contribute to the realization of durable solutions for IDPs. Planned appropriately, they can foster social, economic, and political integration of IDPs into the local communities. Energy projects can also shift mentalities from war to peace. The economic implications of adequate energy supply also cannot be overlooked.

Acknowledgments

The authors received support from the Institute for Future Initiatives (University of Tokyo) and the research grant JSPS Kaken 20KK0034. We would also like to express our gratitude to the experts whom we interviewed and the participants in the focus groups and to UNIMINUTO for the coordination of these.

References

Agudelo Blandón, E. N. (2015). Prácticas de no violencia: Interculturalidad y poétias para curar rabia. In A. Castillejo Cuéllar, E. A. Rueda Barrera, E. N. Agudelo Blandón, & N. Quinceno Toro (Eds.), *Proceso de paz y perspectivas democráticas en Colombia*. CLACSO.

Arango-Aramburo, S., Turner, S. W. D., Daenzer, K., Ríos-Ocampo, J. P., Hejazi, M. I., Kober, T., . . . van der Zwaan, B. (2019). Climate impacts on hydropower in Colombia: A multi-model assessment of power sector adaptation pathways. *Energy Policy, 128*(July 2018), 179–188. Available from https://doi.org/10.1016/j.enpol.2018.12.057.

Arbelaez, M. A., Garcia, F., & Sandoval, C. (2007). El crédito no bancario: una alternativa para la bancarización y la reducción de la pobreza. *El caso del "Crédito Fácil para Todo" de CODENSA*. Available from http://hdl.handle.net/11445/1743.

Asmann, P. (2019). Battles between former FARC groups displace hundreds in Colombia. *InSight Crime*. Available from https://www.insightcrime.org/news/analysis/colombia-forced-displacement-farc/.

Baker, J. L. (2012). *Climate change, disaster risk, and the urban poor: Cities building resilience for a changing world*. Available from https://doi.org/10.1596/978-0-8213-8845-7.

Baldwin, E., Brass, J. N., Carley, S., & Maclean, L. M. (2015). Electrification and rural development: Issues of scale in distributed generation. *WIREs Energy and Environment, 4*(2), 196–211. Available from https://doi.org/10.1002/wene.129.

Barrucand, M. G., Giraldo Vieira, C., & Canziani, P. O. (2017). Climate change and its impacts: Perception and adaptation in rural areas of Manizales, Colombia. *Climate and Development, 9*(5), 415–427. Available from https://doi.org/10.1080/17565529.2016.1167661.

Bhatia, M., & Angelou, N. (2015). Beyond Connections Energy Access Redefined. 58–177. http://www.worldbank.org/content/dam/Worldbank/Topics/EnergyandExtract/Beyond_Connections_Energy_Access_Redefined_Exec_ESMAP_2015.pdf.

Bhattacharyya, S. C. (2012). Energy for sustainable development energy access programmes and sustainable development: A critical review and analysis. *Energy for Sustainable Development, 16*(3), 260–271. Available from https://doi.org/10.1016/j.esd.2012.05.002.

Bocarejo, J. P., Portilla, I. J., Velásquez, J. M., Cruz, M. N., Peña, A., & Oviedo, R. (2014). An innovative transit system and its impact on low income users: The case of the Metrocable in Medellín. *Journal of Transport Geography, 39*(July 2014), 49–61. Available from https://www.sciencedirect.com/science/article/abs/pii/S096669231400132X.

Brand, P., & Dávila, J. D. (2011). Mobility innovation at the urban margins. *City, 15*(6), 647–661. Available from https://doi.org/10.1080/13604813.2011.609007.

Brass, J. N., Carley, S., MacLean, L. M., & Baldwin, E. (2012). Power for development: A review of distributed generation projects in the developing world. *Annual Review of Environment and Resources, 37*(1), 107–136. Available from https://doi.org/10.1146/annurev-environ-051112-111930.

Burke, M. J., & Stephens, J. C. (2017). Energy democracy: Goals and policy instruments for sociotechnical transitions. *Energy Research and Social Science, 33*(September), 35–48. Available from https://doi.org/10.1016/j.erss.2017.09.024.

Cabraal, R. A., Barnes, D. F., & Agarwal, S. G. (2005). Productive uses of energy for rural development. *Annual Review of Environment and Resources, 30*(1), 117–144. Available from https://doi.org/10.1146/annurev.energy.30.050504.144228.

Calle, H. (2018). La energía limpia que los excombatientes llevaron a Miravalle. *El Espectador*. Available from https://www.elespectador.com/noticias/medio-ambiente/la-energia-limpia-que-los-excombatientes-llevaron-miravalle-articulo-826191.

Centro Nacional de Memoria Historica. (2015). *Una nación desplazada: Informe nacional del desplazamiento forzado en Colombia*. Available from http://www.centrodememoriahistorica.gov.co/descargas/informes2015/nacion-desplazada/una-nacion-desplazada.pdf.

Chaurey, A., Krithika, P. R., Palit, D., Rakesh, S., & Sovacool, B. K. (2012). New partnerships and business models for facilitating energy access. *Energy Policy, 47*((Suppl. 1), 48–55. Available from https://doi.org/10.1016/j.enpol.2012.03.031.

Colombia2020. (2018). No solo subsidios, los desplazados necesitan legalizar sus asentamientos: Corte constitucional. *El Espectador*. Available from https://www.elespectador.com/colombia2020/pais/no-solo-subsidios-los-desplazados-necesitan-legalizar-sus-asentamientos-corte-constitucional-articulo-857067.

Cosoy, N. (2016). ¿Por qué empezó y qué pasó en la guerra de más de 50 años que desangró a Colombia? *BBC News*. Available from https://www.bbc.com/mundo/noticias-america-latina-37181413.

de Bercegol, R., & Monstadt, J. (2018). The Kenya slum electrification program. Local politics of electricity networks in Kibera. *Energy Research and Social Science, 41*(August 2017), 249–258. Available from https://doi.org/10.1016/j.erss.2018.04.007.

DNP. (2016). *Caracterizacion territorios con programas de desarrollo con enfoque territorial-PDET*. Available from https://colaboracion.dnp.gov.co/CDT/PolticadeVctimas/ConstrucciióndePaz/CaracterizaciónPDET.pdf.

Dube, O., & Vargas, J. F. (2013). Commodity price shocks and civil conflict: Evidence from Colombia. *Review of Economic Studies, 80*(4), 1384–1421. Available from https://doi.org/10.1093/restud/rdt009.

Econometria. (2016). *External Evaluation of the Joint UNHCR – UNDP Program "Transitional Solutions Initiative - TSI"* (Issue October). http://tsicolombia.org/sites/acnur/files/descargas/executive_summary_tsi_evaluation_colombia.pdf.

El Espectador. (2018). Responsabilidades y dudas que deja la crisis de Hidroituango. *El Espectador*. Available from https://www.elespectador.com/noticias/investigacion/responsabilidades-y-dudas-que-deja-la-crisis-de-hidroituango-articulo-789448.

Fondo Europeo para la Paz en Colombia. n.d. *Sobre el Fondo Europeo para la Paz en Colombia*. https://www.fondoeuropeoparalapaz.eu/sobre-el-fondo/

Forero, S. (2019). ¿Cuál es la importancia de los Programas de Desarrollo con Enfoque Territorial (PDET)? *El Espectador*. Available from https://www.elespectador.com/colombia2020/territorio/cual-es-la-importancia-de-los-programas-de-desarrollo-con-enfoque-territorial-pdet-articulo-857704.

GMH. (2013). ¡BASTA YA! Colombia: Memorias de guerra y dignidad Informe General Grupo de Memoria Histórica. In Centro Nacional de Memoria Histórica. Centro Nacional de Memoria Historica. https://doi.org/10.1017/CBO97811074153240.004.

Gonzalez Gonzalez, F. E. (2014). *Poder y violencia en Colombia*. CINEP/PPP.

Gottwald, M. (2016). Peace in Colombia and solutions for its displaced people. *Forced Migration Review, 52*(May), 14–17.
Goulet, D. (1978). *Sufficiency for all: The basic mandate of development and social economics, . Review of Social Economy* (Vol. 36, pp. 243–261). Taylor & Francis, Ltd. Available from https://doi.org/10.2307/29768935.
Graviano, N., Götzelmann, A., Nozarian, N., & Wadud, J. (2017). *Towards an integrated approach to reintegration in the context of return.*
Guevara, J. P. (2015). El Plan Colombia o el desarrollo como seguridad. *Revista Colombiana de Sociología, 38*(1), 63–82. Available from https://doi.org/10.15446/rcs.v38n1.53264.
Harild, N. (2016). Forced displacement: A development issue with humanitarian elements. May, 2014–2017.
Harild, N., & Christensen, A. (2010). The Development Challenge of Finding Durable Solutions for Refugees and Internally Displaced People. *World Development Report 2011* Background Note, 1–8. http://web.worldbank.org/archive/website01306/web/pdf/wdrbackgroundnoteondisplacement_04dbd.pdf?keepThis = true&TB_iframe = true&height = 600&width = 800.
Hernandez, E. (2019, January 17). En Colombia también hay huachicoleo, usan combustibles para elaborar cocaína. El Sol de México. https://www.elsoldemexico.com.mx/mexico/justicia/huachicol-gasolina-ecopetrol-colombia-cocaina-mexico-robo-combustibles-2933692.html.
IASC. (2010). *IASC framework on durable solutions for interllay displaced persons.* Available from https://www.unhcr.org/50f94cd49.pdf.
IDEAM. (2017). *Análisis cambio climático en Colombia: Tercera Comunicación Nacional.*
IOM. (2017). *IOM framework for addressing internal displacement.*
Kalin, W. (2005). The guiding principles on internal displacement as international minimum standard and protection tool. *Refugee Survey Quarterly, 24*(3), 27–36. Available from https://doi.org/10.1093/rsq/hdi050.
Kenner, D. (2017). Solar and wind energy: Driver of conflict or force for peace? Why green economy?
Komives, K., Foster, V., Halpern, J., & Wodon, Q. (2005). *Water, electricity, and the poor: Who benefits from utility subsidies?* . Available from http://documents.worldbank.org/curated/en/606521468136796984/Water-electricity-and-the-poor-who-benefits-from-utility-subsidies.
Koser, K. (2007). Addressing internal displacement in peace processes, peace agreements and peace-building (Issue September).
Kälin, W. (2003). The legal dimension. *Forced Migration Review, 17,* 1–10. Available from https://doi.org/10.1080/01419870.2012.634507.
Lehne, J., Blyth, W., Lahn, G., Bazilian, M., & Grafham, O. (2016). Energy services for refugees and displaced people. *Energy Strategy Reviews, 13–14,* 134–146. Available from https://doi.org/10.1016/j.esr.2016.08.008.
Leon, C. (2016). *Analysis of Utility Models for the Base of the Pyramid* (Issue February). https://static1.squarespace.com/static/5a168fa5f6576eb8bfa0a5e5/t/5a340e3041920241eb51ce23/1513360960024/UTILITY + BUSINESS + MODELS + FOR + THE + BASE + OF + THE + PYRAMID.
Ley 1448 de 2011, (2011) (testimony of Colombian Congress). https://www.unidadvictimas.gov.co/sites/default/files/documentosbiblioteca/ley-1448-de-2011.pdf
Listo, R. (2018). Gender myths in energy poverty literature: A critical discourse analysis. *Energy Research and Social Science, 38*(January), 9–18. Available from https://doi.org/10.1016/j.erss.2018.01.010.

Mahat, I. (2011). Gender, energy, and empowerment: A case study of the rural energy development program in Nepal. *Development in Practice, 21*(3), 405–420. Available from https://doi.org/10.1080/09614524.2011.558062.

Marijnen, E., & Schouten, P. (2019). Electrifying the green peace? Electrification, conservation and conflict in Eastern Congo. *Conflict, Security & Development, 19*(1), 15–34. Available from https://doi.org/10.1080/14678802.2019.1561615.

Martínez Ortiz, A., Afanador, E., Zapata, J. G., Núñez, J., Ramírez, R., Yepes, T., & Garzón, J. C. (2013). Análisis de la situación energética de Bogotá y Cundinamarca. *Cuadernos de Fedesarrollo, 45*, 1–330.

McDermott, J. (2004). Colombia cracks down on oil theft. *BBC News*. Available from http://news.bbc.co.uk/2/hi/americas/3489829.stm.

Mozersky, B. D., & Kammen, D. (2018). *Thinking outside the box in South Sudan: How renewable energy can serve as a building block for peace*. Available from https://www.usip.org/sites/default/files/2018-01/sr418-south-sudans-renewable-energy-potential-a-building-block-for-peace.pdf.

Parikh, P., Chaturvedi, S., & George, G. (2012). Empowering change: The effects of energy provision on individual aspirations in slum communities. *Energy Policy, 50*, 477–485. Available from https://doi.org/10.1016/j.enpol.2012.07.046.

Pizarro, E. (1987). *La guerrilla en Colombia. Entre la guerra y la paz. Puntos de vista sobre la crisis de los años 80* (No. 141; Controversia).

Portilla Benavides, A. C., & Correa, C. (2015). *Estudio sobre la implementacion del Programa de Reparacion Individual en Colombia*. Available from https://www.ictj.org/sites/default/files/ICTJ-COL-Estudio-reparacion-individual-2015.pdf.

Reuters. (2018, December). Colombia's Cano Limon pipeline halted by bomb attack. Reuters. https://www.reuters.com/article/us-colombia-ecopetrol/colombias-cano-limon-pipeline-halted-by-bomb-attack-idUSKBN1KX01V.

Rysankova, D., Portale, E., & Carletto, G. (2016). Measuring energy access introduction to the multi-tier framework SE4ALL knowledge hub publications. *Sustainable Energy for All*. Available from https://www.seforall.org/sites/default/files/MTFpresentation_SE4ALL_April5.PDF.

Sabogal, J. (2019). ¿Qué ha pasado con la legalización de barrios en Bogotá? *RCN Radio*. Available from https://www.rcnradio.com/bogota/que-ha-pasado-con-la-legalizacion-de-barrios-en-bogota.

Samad, H., & Zhang, F. (2019). *Electrification and Women's Empowerment Evidence from Rural India* (No. 8796; Policy Research Working Paper, Issue March). https://ssrn.com/abstract = 3362640.

Samuel Hall., NRC., & IDMC. (2018). Escaping war: Where to next? A research study on the challenges of IDP protection in Afghanistan. Available from http://www.internal-displacement.org/assets/publications/2018/20180124-NRC-IDMC-SamuelHall-escaping-war-where-to-next.pdf.

Secretaria de Planeacion y Ordenamiento Territorial. (2017). *Plan de Ordenamiento Territorial (POT) Soacha, Cundinamarca - Documento y cartografia de diagnostico terrirotorial urbano y rural*. Available from http://www.alcaldiasoacha.gov.co/phocadownloadpap/secretaria_de_planeacion/POT/DOCUMENTODIAGNOSTICOFINAL-2018.pdf.

Secretaría Distrital de Hábitat. (2017). *Determinantes de la tenencia formal de vivienda en Bogotá - Encuesta multipropósito Bogotá*. Available from https://habitatencifras.habitatbogota.gov.co/documentos/Estudios_Sectoriales/Tenencia.pdf.

Semana. (2002). El nuevo narcotrafico. *Semana*. Available from https://www.semana.com/nacion/articulo/el-nuevo-narcotrafico/54190-3.

Siciliano, G., Urban, F., Tan-Mullins, M., & Mohan, G. (2018). Large dams, energy justice and the divergence between international, national and local developmental needs and priorities in the global South. *Energy Research and Social Science, 41*(March), 199–209. Available from https://doi.org/10.1016/j.erss.2018.03.029.

Souter, J. (2014). Durable solutions as reparation for the unjust harms of displacement: Who owes what to refugees? *Journal of Refugee Studies, 27*(2), 171–190. Available from https://doi.org/10.1093/jrs/fet027.

Sovacool, B. K. (2013). Expanding renewable energy access with pro-poor public private partnerships in the developing world. *Energy Strategy Reviews, 1*(3), 181–192. Available from https://doi.org/10.1016/j.esr.2012.11.003.

Sovacool, B. K., & Dworkin, M. H. (2015). Energy justice: Conceptual insights and practical applications. *Applied Energy, 142*, 435–444. Available from https://doi.org/10.1016/j.apenergy.2015.01.002.

Stokes, D. (2005). *America's other war: Terrorizing Colombia.* Zed Books.

Superservicios. (2017). *Diagnostico de la calidad del servicio de energia electrica en Colombia* (Issue May).

TECHO. (2015). Derecho a Bogotá. https://issuu.com/techocolombia/docs/derecho_a_bogot.

Torres, M. A., Caballero, J. H., & Awad, G. (2016). Social and environmental impactsof hydroelectric projects case study: Hidroelectrica Ituango. *Iberoamerican Journal of Program Management, 7*(1), 94–115.

Torres Tovar, C. A. (2009). *Ciudad informal colombiana: Barrios construidos por la gente.* Universidad Nacional de Colombia. Available from http://www.facartes.unal.edu.co/fa/institutos/ihct/publicaciones/ciudad_informal.pdf.

Torres Tovar, C. A. (2012). Legalización de barrios: acción de mejora o mecanismo de viabilización fiscal de la ciudad dual. *Bulletin de l'Institut Français d'études Andines, 41*(3), 441–471. Available from https://doi.org/10.4000/bifea.304.

UNHCR. (2018). *Statistics and operational data.* Available from https://www.unhcr.org/figures-at-a-glance.html.

Unidad para la Atención y la Reparación Integral a las Víctimas. (2019). *Corea desarrollará nuevos proyectos con la Unidad en beneficio de las víctimas.* Available from https://www.unidadvictimas.gov.co/es/cooperacion-internacional/corea-desarrollara-nuevos-proyectos-con-la-unidad-en-beneficio-de-las.

UNSG. (2011). *Protection of and assistance to internally displaced persons.* Available from https://www.ohchr.org/Documents/Issues/IDPersons/A-66-285.pdf.

UPME. (2014). *Plan Indicativo de Expansión de Cobertura 2013–2017.*

Valdés Correa, B. (2017). La población flotante de Tumaco. Available from https://www.elespectador.com/colombia-20/paz-y-memoria/la-poblacion-flotante-de-tumaco-article/.

World Bank. (2017). *Forcibly displaced: Toward a development approach supporting refugees, the internally displaced, and their hosts.* Available from https://doi.org/10.1596/978-1-4648-0938-5.

5

Possible role of renewables in the Myanmar peace process

Masako Numata[1,2] and Masahiro Sugiyama[3]

[1]International Projects Division, The Nippon Foundation, Tokyo, Japan [2]MbSC2030, The University of Tokyo, Tokyo, Japan [3]Institute for the Future Initiative, The University of Tokyo, Tokyo, Japan

Chapter outline

Energy Policy for Peace. DOI: https://doi.org/10.1016/B978-0-12-817350-3.00005-5

5.1 Electrification in Myanmar

5.1.1 Energy access in Myanmar

5.1.1.1 The overall situation in Myanmar

The rate of electrification in Myanmar has not yet surpassed 50%, which is low by global standards (Billen & Bianchi, 2019). The Myanmar government has set a goal of 100% electrification by 2030 (Ministry of Electricity & Energy Electricity Supply Enterprise The Republic of the Union of Myanmar, 2019).[1,2] One challenge that is recognized as requiring a solution is the necessity to electrify peripheral regions with (previous) ethnic conflicts in order to reach 100% electrification. Myanmar has ratified the Paris Agreement (The Republic of the Union of Myanmar, 2015), so it is preferable that electrification occurs through renewal energy sources rather than through the use of fossil fuels. We analyzed the cost competitiveness of renewable energy minigrids, which have gained attention as off-grid methods of electrification, and diesel generators (Numata, Sugiyama, Mogi, Wunna, & Anbumozhi, 2018) as well as barriers and other factors that are unrelated to the economy (Numata, Sugiyama, & Mogi, 2018).

However, as Sovacool, Heffron, McCauley, and Goldthau (2016) have pointed out, "All too often, energy policy and technology discussions are limited to the domains of engineering and economics." The analysis of the energy sector in Myanmar until now has been focused on the technical side and on economic evaluations that are primarily concerned with cost. Sovacool (2012) stated separately that "social, political, and cultural domains" also need to be considered and understood.

[1]The vast majority of the text of this chapter was drafted before February 2021, during which the National League for Democracy was heading the government of Myanmar. The political turmoil in February and since then is not reflected in the present chapter.
[2]Some of the contents of this chapter have already been published. See Numata, Sugiyama, and Mogi (2021).

This point is quite relevant to Myanmar. As the country makes progress on electrification, it increasingly faces the tough question of how to electrify peripheral regions with past/existing ethnic conflicts. In the past, development projects in those regions, including some hydropower dams, suffered from what is called energy injustice. In recent years, the declining costs of distributed renewables began changing the energy landscape, perhaps providing an opportunity for rural villages to take advantage of new technologies to gain access to modern electricity services.

This chapter examines the issue of electrification with renewables, especially in the context of the effects of electrifying rural areas with ethnic conflicts. Our point of departure is that distributed renewables can play a role in electrifying rural areas with past and/or ongoing ethnic conflicts and that such a contribution could possibly help to address the ethnic conflicts. To do this, the chapter takes a staged approach. First, a brief historical review of ethnic conflicts since around the World War Two is provided. Second, the chapter critically appraises development and electrification projects in conflict areas. Third, this chapter presents the initial findings from semistructured interviews with key stakeholders, focusing on high-level messages.

5.2 Brief overview of conflicts in Myanmar

5.2.1 Ethnic groups

Since gaining independence in 1948 Myanmar has been witnessing a series of primarily ethnic conflicts in what may be described as one of the world's longest-running civil wars. The population is approximately two-thirds Bamar, with the remainder made up of various ethnic minorities. The government of Myanmar publicly recognizes 135 ethnic groups (Central Intelligence Agency, 2019; Lall & South, 2018). Table 5.1 shows the major national ethnic groups in Myanmar and their constituent subgroups.

Although data related to religion and ethnicity were not collected in the 2014 census in Myanmar, religious data were published in 2016 (Department of Population Ministry of Immigration & Population, 2015). However, data on ethnicity have not been published (Tun, 2017; S. Y. Aung, 2018). Roughly 100 languages are spoken within Myanmar; the languages are sometimes synonymous with ethnicity and sometimes not (Ethnologue, n.d.). Languages with their own alphabets as well as languages without a specific alphabet also exist in Myanmar (Everson & Hosken, 2006). In Chin State, so many languages exist that the residents of

Table 5.1 List of ethnic groups.[3]

Major national ethnic groups	Number of subgroups	Subgroups
Kachin	12	Kachin, Trone, Dalaung, Jinghpaw, Guari, Hkahku, Duleng, Maru (Lawgore), Rawang, Lashi (La Chit), Atsi, Lisu
Kayah (Karenni)	9	Kayah, Zayein, Ka-Yun (Padaung), Gheko, Kebar, Bre (Ka-Yaw), Manu Manaw, Yin Talai, Yin Baw
Kayin (Karen)	11	Kayin, Kayinpyu, Pa-Le-Chi, Mon Kayin (Sarpyu), Sgaw, Ta-Lay-Pwa, Paku, Bwe, Monnepwa, Monpwa, Shu (Pwo)
Chin	53	Chin, Meithei (Kathe), Saline, Ka-Lin -Kaw (Lushay), Khami, Awa Khami, Khawno, Kaungso, Kaung Saing Chin, Kwelshin, Kwangli (Sim), Gunte (Lyente), Gwete, Ngorn, Zizan, Sentang, Saing Zan, Za-How, Zotung, Zo-Pe, Zo, Zah nyet (Zanniet), Tapong, Tiddim (Hai-Dim), Tay-Zan, Taishon, Thado, Torr, Dim, Dai (Yindu), Naga, Tangh kul, Malin, Panun, Magun, Matu, Miram (Mara), Mi-er, Mgan, Lushei (Lushay), Laymyo, Lyente, Lawhtu, Lai, Lai zao, Wakim (Mro), Haulngo, Anu, Anu n, Oo-Pu, Lhinbu, Asho (Plain), Rongtu
Bamar	9	Bamar, Dawei, Beik, Yaw, Yabein, Kadu, Ganan, Salon, Hpon
Mon	1	Mon
Rakhine	7	Rakhine, Kamein, Kwe Myi, Daingnet, Marama gyi, Mro, Thet
Shan	33	Shan, Yun (Lao), Kwi, Pyin, Yao, Danaw, Pale, En, Son, Khamu, Kaw (Akha-E-Kaw), Kokang, Khamti Shan, Hkun, Taung yo, Danu, Palaung, Man Zi, Yin Kya, Yin Net, Shan Gale, Shan Gyi, Lahu, Intha, Eik-swair, Pa-O, Tai-Loi, Tai-Lem, Tai-Lon, Tai-Lay, Maingtha, Maw Shan, Wa

Source: From Smith, M. (1994). Ethnic groups in Burma. (A.-M. Sharman, Ed.), Anti-Slavery international. London: Anti-Slavery International. <http://www.ibiblio.org/obl/docs3/Ethnic_Groups_in_Burma-ocr.pdf>. Embassy of the Union of Myanmar Brussels (n.d.).

villages on opposite sides of a valley may speak different dialects (Takahashi, 2018). The official language of Myanmar is Burmese, the language of the Bamar people, the country's principal ethnic group (Smith, 1994).

The administrative districts of Myanmar are made up of seven states and seven regions as well as self-administered zones or divisions. Most of the population of the seven regions is Bamar. The seven states are Chin, Kachin, Kayin, Kayah, Mon, Rakhine, and Shan; each is named for the ethnic group that makes up the

[3]The Rohingya are not included in the list of 135 ethnicities recognized by the government. Because a discussion of the Rohingya would be extensive and complex, it is not included in the discussion in this chapter.

majority of its population. However, the distribution of ethnicities within each state is diverse and does not necessarily coincide with the ethnolinguistic boundaries of states, regions, or townships. For example, a number of Kayin (Karen) people and Rakhine people reside in the Ayeyarwady region, and the ministers of Kayin Ethnic Affairs and Rakhine Ethnic Affairs both belong to the Ayeyarwady Region Government.

According the Myanmar Constitution, enacted in 2008 there are six self-administered zones or divisions within the country: the Danu Self-Administered Zone, the Kokang Self-Administered Zone, the Pa'O Self-Administered Zone, the Pa Laung Self-Administered Zone, the Wa Self-Administered Division in Shan State, and the Naga Self-Administered Zone in Sagaing Region. These zones are self-administered by the ethnic groups for which they are named (Office of the Civil Service Commission, n.d.).

5.2.2 Ethnic armed organizations

Myanmar has one of the longest-running conflicts in the world. The war has become entangled with the history of neighboring countries and thus has become much more complicated over time. There are approximately 20 ethnic armed organizations (EAOs) of various sizes, degrees of sophistication, and origin. The more influential EAOs have organized initiatives for education, health, and so on, and they provide social services to the area that they control independently or through coordination with civil society organizations (CSOs) (Christophersen & Stave, 2018). For example, the leadership of one of the influential EAOs, the Karen National Union (KNU), is determined by formal election (Naing, 2017). Table 5.2 lists the names and acronyms of the major EAOs, along with the names of their armed wings. Table 5.3 presents a short history of Myanmar.

Table 5.3 summarizes the modern history of Myanmar.

In discussing the areas where minorities live in Myanmar, designations are often used that sort the regions into black, brown (or gray), and white areas. Areas that are controlled by the state or regional government or by the Myanmar army ("Tatmadaw") are designated as "white" areas, areas controlled by EAOs are "black" areas, and mixed control areas are referred to as "brown" or "gray" areas. Access to black areas is restricted for foreigners, and there are no maps that show the detailed locations of villages. Access to brown/gray areas must be arranged in advance by foreign visitors, and as these areas are controlled by two forces, they are said to be the hardest to live in even for local villagers, as taxes are collected

Table 5.2 Names and abbreviation of active ethnic armed organizations and their armed wings.

Name of organization	Name of armed wing
ULA/**AA**	United League of Arakan/Arakan Army
ABSDF	All Burma Students' Democratic Front
ALP/ALA	Arakan Liberation Party/Arakan Liberation Army
CNF/can	Chin National Front/Chin National Army
DKBA (formerly DKBA-5)	Democratic Karen Benovelent (Buddhist) Army
KIO/KIA	Kachin Independence Organization/Army
KNPP/KA	Karenni National Progressive Party/Karenni Army
KNU/KNLA	Karen National Union/Karen National Liberation Army
KNU/KNLA-PC (not related to KNU or KNLA)	Karen National Union/Karen National Liberation Army Peace Council
KUKI	Kuki National Organisation/Kuki National Army
LDU	Lahu Democratic Union
MNDAA	Myanmar National Democratic Alliance Army
PSC/**NDAA**(-ESS)	Peace and Solidarity Committee (Mongla)/National Democratic Alliance Association-East Shan State
NMSP/NMLA	New Mon State Party/Mon National Liberation Army
NSCN-K	National Socialist Council of Nagaland-Khaplang
PNLO/PNLA	Pa-O National Liberation Organisation/Army
RCSS/SSA(-S)	Restoration Council of Shan State/Shan State Army(-South)
SSPP/SSA(-N)	Shan State Progress Party/Shan State Army(-North)
PSLF/**TNLA**	Palaung State Liberation Front/Ta'ang National Liberation Army
UWSP/**UWSA**	United Wa State Party/Army

Note: Many organizations have a political wing and an armed wing, but in this chapter, for ease of comprehension, no distinction is made between the political and armed wings and the acronym in bold face is used.
Source: From Tønnesson, S., Aung, N. L., & Nilsen, M. (2019). Will Myanmar's Northern Alliance Join the Peace Process ? (PRIO POLICY BRIEF). Oslo. <https://www.prio.org/utility/DownloadFile.ashx?id = 1757&type = publicationfile>.; Myanmar Peace Monitor. (2016). Armed ethnic groups 2016. Yangon: Myanmar Peace Monitor, The Burma News International. <http://www.mmpeacemonitor.org/#!armed-ethnic-groups/c1q70>.

by both the government and the EAOs, forcing residents to work for one side or the other (Jolliffe, 2014).

5.2.2.1 Nationwide ceasefire agreement

The central framework for the current Myanmar peace process is the Nationwide Ceasefire Agreement (NCA). In October 2015 the NCA was signed by the Myanmar government under the previous Thein Sein Administration, and with the agreement of eight EAOs. Since 2016, when the National League for Democracy (NLD) Administration won the November 2015 election in a landslide, the

Table 5.3 Brief history of Myanmar.[4]

British colonial period	Direct governance in Burma proper area (Tanintharyi region, Rakhine state, Bago region, and Ayeyarwady region at present) and autonomous governance in peripheral areas. Appointment of ethnic minority for military and police.
1943	One time independence from British colony with Japanese army's support.
1945	A coup d'état is led by Aung San against the Burmese government, a Japanese puppet state then return to the British.
1947	The first Panglong Agreement is concluded but signer General Aung San is assassinated. KNU is formed in Kayin state, southeastern Myanmar.
1948	Independence from Britain with first prime minister U Nu. Instability due to uprisings by minority peoples and the Communist Party of Burma (CPB). KIO is formed in Kachin state, and Shan State Army is formed in Shan state, north Myanmar.
1962	Coup d'état by General Ne Win. Military regime continues until 1988 even after 1981 retirement of Ne Win as president. "Burmese Socialism." Tatmadaw adopts the infamous "four cuts" strategy to cut off coordination between EAOs and villages in food, funds, intelligence, and recruits. Villagers are unable to farm and are forcibly relocated and become internally displaced persons (IDPs).
1988	The 8888 uprising: Demonstrations demanding democracy cause collapse of the socialist administration after 26 years, and the military seizes power (under General Saw Maung). CPB collapses, and each ethnic group forms its own EAO: in northeast Shan State the MNDAA of the Kokang, the UWSA of the Wa, and the NDAA of the Mong La Shan and the Aka; in northeast Kachin state the New Democratic Army-Kachin (NDA-K) of the Kachin people. Future prime minister Khin Nyunt concludes a ceasefire agreement, and these four EAOs are allowed to turn the areas they control into special administrative districts (SADs).
1990	Although the National League for Democracy (NLD) wins the general elections in a landslide, the results are ignored and the military suppresses dissent. Aung San Suu Kyi is put under house arrest (three times until 2010 for a total of 15 years).
1992	Than Shwe becomes head of state. He is commander-in-chief until 2011. He ranks No. 4 in the "World's Worst Dictators" list in 2009. The push for a bilateral ceasefire under Thai government pressure to send back about 10,000 IDP of Mon ethnic group. NMSP and the government of Prime Minister Khin Nyunt conclude the bilateral ceasefire, but the movement wanes with the 2004 overthrow of the Khin Nyunt.
2006	The capital is moved to Nay Pyi Taw.
2007/9	Saffron Revolution (a nationwide demonstration by monks).
2007/10	Thein Sein becomes prime minister.
2008	Cyclone Nargis. Nearly 140,000 people die.
2010	General elections are implemented but boycotted by NLD. USDP wins.

(Continued)

[4]This became the base of the armed wings of the SSPP and RCSS, the Shan State Army-North and Shan State Army-South.

Table 5.3 (Continued)

2011	The Thein Sein administration begins.
2011	Conflicts restart between the national military and the KIO. Intermittent clashes between a number of armed organizations continue in northeastern Myanmar.
2012	Conflicts occur between Buddhists and Muslims in Rakhine State.
2015/10	The Myanmar government and eight EAO agree on a nationwide ceasefire (Nationwide Ceasefire Agreement).
2015/11	NLD wins the general elections in a landslide.
2016	NLD administration begins.
	The Northern Alliance is established by four non-ceasefire EAOs: KIO, AA, TNLA, and MNDAA.
	UWSA (which has a close relationship with China), SSPP, NDAA, and the Northern Alliance members form the Federal Political Negotiation and Consultative Committee. UWSA takes the initiative among nonsignatory EAOs.
	AA moves its activities from Kachin state to Rakhine state. The southern part of Chin state then becomes very active.
2018/2	Two EAOs, namely, NMSP and LDU, sign the NCA.
2018/12	The Tatmadaw announces a four-month unilateral ceasefire that covers Shan state and Kachin state in the northeast part of the country, where the Northern Alliance members are active except for AA in Rakhine state.
2020/10	A general election is held.
2021/2	Political turmoil erupts because of the contest over the election results.

Source: From Myint-U (2011); Kramer (2012); Bi (2012); Ministry of Foreign Affairs of Japan (2017); Burke, Williams, Barron, Jolliffe, and Carr (2017); Tønnesson et al. (2019); Nyein (2018b); Wallechinsky (2009); Kubo (2014).

NCA framework remained in place; it even gained two more EAO signatories in February 2018. EAOs differ considerably in terms of size and legitimacy, and one of these categories of difference is between NCA signatories and nonsignatories. Also, among nonsignatory groups, there are EAOs that have agreed to bilateral ceasefires with the government and EAOs that have not agreed to any ceasefires. Table 5.4 shows the participation status of EAOs in the NCA. Of the approximately 20 major EAOs, roughly half, or 10 EAOs, are signatories. However, if the size of the EAO signatories and nonsignatories is considered, a total number of 66,000 soldiers form the nonsignatory EAOs, including the UWSA, which is thought to have a force upwards of 30,000, while total signatory forces total only 17,000, or one-fourth of total EAO forces (Myanmar Peace Monitor, 2016). Generally speaking, EAOs from the southeastern regions of Myanmar on the Thai border are NCA signatories, while EAOs from regions in the northeast along the Chinese border are nonsignatories (Burma News International, 2017; United States Institute of Peace, 2018). Before and after the

Table 5.4 Participation status of Nationwide Ceasefire Agreement and ethnic armed organizations (Myanmar Peace Monitor, n.d.-b).

NCA signatories	NCA nonsignatories	
	Bilateral ceasefire	Non-ceasefire
RCSS	UWSA	KIO
KNU	SSPP	TNLA
DKBA	NDAA	AA
ABSDF	KNPP	MNDAA
PNLO	NSCN-K	KUKI
CNF		
KNU/KNLA-PC		
ALP		
NMSP (signed in 2018 Feb.)		
LDU (signed in 2018 Feb.)		

Source: From Myanmar Peace Monitor. Stakeholders overview. (n.d.-b). <http://mmpeacemonitor.org/stakeholders/stakeholders-overview> Accessed 25.04.19.

enactment of the NCA, the occurrence of IDPs shifted from southern Shan State, Kayah State, Kayin State, and the southeast Tanintharyi region in 2006 to regroup in Kachin State, Shan State, Rakhine State, and a portion of Mon State in 2016 (Burke et al., 2017). Unfortunately, the number of battles fought within EAO states has not decreased in Myanmar since the signing of the NCA (Bynum, 2018).

The EAOs that are considered to be the main NCA signatories are the KNU, the Restoration Council of Shan State (RCSS), and the New Mon State Party (NMSP), which became a signatory EAO in 2018 (Tønnesson et al., 2019). Also, the origins of the All Burma Students' Democratic Front (ABSDF), which began as a student movement, have indirectly had an impact on higher education in Myanmar.

The origins of the main EAOs, divided into NCA signatories and nonsignatories, are described in the appendix at the end of the chapter.

5.2.2.2 Conflicts between ethnic armed organizations

Although the relationships between each of the EAOs and the Myanmar government or Tatmadaw have been different, relations between EAOs are certainly not limited to their alliances. EAOs are often one another's enemies. The Restoration Council of Shan State (RCSS), an NCA signatory, and the Shan State Progress Party (SSPP), an NCA nonsignatory, are both active in Shan State. Although the RCSS is considered to be active in southern Shan State while the SSPP is considered to be active in northern Shan State, they do not keep entirely to themselves. Since the end of 2018, combat has frequently broken out in northern Shan State, with more than 2000 people becoming internally displaced (Weng, 2018; Weng, 2019b). As the Shan people did not welcome this conflict (Tønnesson et al., 2019), peace talks finally began between the two rival EAOs in 2019 (Asianews.it, 2019). However, the Ta'ang National Liberation Army (TNLA), which allied with the SSPP against the RCSS, has not been included in the peace talks (Weng, 2019b). Conflicts between other groups have occurred, such as those between the RCSS and Pa-O National Liberation Organisation (PNLO) and between the KNU and NMSP in 2018 to April 2019 (Bynum, 2019) and those between the TNLA and RCSS, United Wa State Army (UWSA) and National Democratic Alliance Association (NDAA), and Arakan Army (AA) and Arakan Liberation Party (ALP) in 2015–16 (Burma News International, 2017).

5.2.3 Issues with the current peace process

When the NLD, led by current State Counselor Daw Aung San Suu Kyi, who was a de facto leader of the government, took over the power, people hoped that the peace process would make further progress. However, the peace process has stalled, as evidenced by the agreement of only two EAO signatories since the NLD took over.

Although the NCA contains the words "ceasefire agreement," it is not a merely an agreement to cease armed conflict, as the name suggests. NCA is more regarded as "a curiously half-baked seven-chapter draft roadmap for arriving at a comprehensive political agreement" (Tønnesson & Nilsen, 2018). It is not a simple ceasefire agreement; rather, it is a roadmap for a national reconciliation. The roadmap is as follows:

1. NCA signing,
2. Drafting and adopting the political dialog,
3. Holding political dialog and negotiating security reintegration,
4. holding Union Peace Conference,
5. signing the Union Accord,
6. ratification by Union parliament, and
7. Implementation (Institute for Security & Development Policy, 2015; The Nationwide Ceasefire Agreement Between

the Government of the Republic of the Union of Myanmar & the Ethnic Armed Organisations, 2015).

Therefore signing the NCA also means agreeing to the roadmap—which contains negotiations for security reintegration—making signing even more difficult for EAOs (Jolliffe, Bainbridge, & Campbell, 2017). There is no other official way to participate in a peace process in Myanmar outside of signing an NCA (Tønnesson & Nilsen, 2018). This exclusion of nonsignatories has allowed a Northern Alliance (see Table 5.3) to emerge and caused the expansion of Arakan Army, a Rakhine insurgent group in Myanmar (Johnson (anonym), personal communication, February 12, 2019).

Distrust between the EAOs and the central Burmese government runs deep. EAOs are generally concerned with the idea that Tatmadaw may strengthen preparedness for action by using the ceasefire period if the EAO signs the NCA (Davis (anonym) of an international CSO, personal communication, December 5, 2018). Regions that have signed the NCA cannot be considered to be in a "postconflict" state (Saferworld & Karen Peace Support Network, 2019; South et al., 2018), and are instead considered to be "fragile" states.

Myanmar has changed to a democratic government by the ruling party NLD, but Tatmadaw continues to hold significant power. The military holds the appointing power for ministers within the Ministries of Defence, Home Affairs, and Border Affairs (Republic of the Union of Myanmar, 2008) Twenty-five percent of seats in the national parliament are assigned to the military (Batcheler, 2018; Republic of the Union of Myanmar, 2008). The ruling party during the former Thein Sein regime was a large promilitary party established under former military administration (Kudo, 2010), and the government and the military coordinated relatively well. However, coordination between the current NLD administration and Tatmadaw cannot be considered to be positive, and the government has been labeled a "two-headed government" (Tønnesson & Nilsen, 2018) led by Aung Sun Suu Kyi, the de facto political leader, and the Myanmar army commander. NLD had confronted the military to demand democratization prior to the change in ruling party, and it is not hard to imagine that close coordination among former warring factions would be difficult. As a result, however, ethnic minorities are forced to negotiate with both the military and the NLD administration, which increases the difficulty of negotiations (Tønnesson & Nilsen, 2018). Compared to the previous regime, which was putting effort into building trust with EAOs, including investment in human resources, some argue that the NLD administration is not doing enough to foster a trusting relationship (South et al., 2018).

Others observe that the NLD administration is more focused on placing the CSOs under the control of the government than previous administrations were (Lall & South, 2018). With regard to government access by environmental NGOs, these organizations generally had "good" access under the Thein Sein administration but have had "moderate" access under the current NLD administration (Simpson & Smits, 2018).

After the change in government, the NLD administration replaced the governmental institution to lead the peace process from the Myanmar Peace Center (MPC) established under Thein Sein administration to the National Reconciliation and Peace Center (NRPC). MPC had sufficient human resources, received financial aid from foreign donors, and functioned well as a platform for the peace process. EAOs and military personnel were also able to converse directly at MPC. State Counselor Daw Aung San Suu Kyi assumed the role of chairperson for NRPC herself. The NRPC has been accused of failing to serve as the coordinator, a role that MPC was able to fulfill (Smith (anonym), personal communication, February 12, 2019).

The NLD government is sometimes accused of being "Burmese-centered" (Shida, 2017). Under the current constitution, the chief minister of the state can be appointed by the central government from members of the state parliament and does not need to be in the majority party within the parliament (Batcheler, 2018). In Rakhine State, the parliament majority is held by the Arakan National Party, but the chief minister is appointed from the NLD party members. Such actions are seen as a marginalization of ethnic minorities by majority members and encourages support for the Arakan Army (AA) in Rakhine State (Weng, 2019a). While General Aung San is widely respected as the hero of independence, the government is leading initiatives to build his statue and name bridges after him in lands of ethnic minorities that do not have any particular connection to him, causing a backlash from these communities (Lynn, 2019; K. H. Aung, 2019).

Not only has the trust not been fostered between the military and EAOs, the trust has also not been built between the civilian government and EAOs. The government claims that poverty reduction through economic development leads to peace, but there are doubts about whether this is as simple as is claimed.

5.3 Development and electrification programs in conflict areas in Myanmar

5.3.1 Sensitivity to development programs in ethnic minority areas due to past conflict

The narrative that raising the standard of living in an area through economic development and a richer livelihood will discourage young people from becoming radicalized and will lead to regional stability and large-scale development projects is often widely celebrated. However, no clear correlation has been observed between development and subnational conflicts, even in countries that have developed rapidly, such as Indonesia, the Philippines, and Thailand (Parks, Colletta, & Oppenheim, 2013). The government in Myanmar is touting economic development as a priority that can lead to the development of peace, but the relationship between development and peace building is complex and has not necessarily had the intended results (Christophersen & Stave, 2018). Economic development can even encourage conflict in some cases (Burke et al., 2017), and implementation of development programs that ignore the social and political context can trigger adverse impacts (South et al., 2018). For ethnic minorities in Myanmar the word "development" often has a negative connotation or association with the government, military, corruption among the EAO leaders, collusion between the government and businesses, corrupt politics, and bribery (Burke et al., 2017). Therefore development programs that are led by the government are often seen with distrust (Christophersen & Stave, 2018; South et al., 2018).

5.3.1.1 Large-scale hydropower

For many people, large-scale hydropower generation is the first image that is evoked by words such as "development" and "energy." Myanmar has rich hydropower resources (ADB, 2016), and the country depends on hydropower for 56% of its generated electricity (Du Pont, 2019). Because of the United Nations SDGs, sustainability is now being emphasized in development projects, and hydropower generation projects often require an assessment of environmental and social impacts at the river basin level along with a conventional assessment of individual projects to ensure that hydropower development is sustainable. International Finance Corporation

(2018) has released a Strategic Environmental Assessment (SEA) of the Myanmar hydropower sector, but the MOEE logo that was found in the draft is deleted in the final report. The Ministry of Natural Resources and Environmental Conservation (MONREC) is the only Burmese ministry listed in the report, while MOEE is reported to be working on a white paper on hydroelectric policy with the Chinese government (Kean, 2019). It cannot be said that the idea of sustainable development of hydropower generation is widely accepted in Myanmar. Kittner and Yamaguchi (2017) noted the need to expand transparency and local engagement in large-scale dam development in Myanmar as well as the need for the international community to provide technical assistance.

Hydropower resources are unevenly distributed in ethnic minority areas in Myanmar. Among the 104 dams (including suspended and identified dams) listed in the hydropower database (IFC, 2017), 28 dams are found in seven regions where Bamar is the majority, while the other 76 are found in seven states. For this reason, large-scale dam development has come to symbolize the exploitation of ethnic minority resources, which makes dam development a difficult project to implement. Development of large-scale dams with large environmental and social impacts to the local communities, such as the Myitsone Dam, were decided under the military regime without any communication with the local community and were completely lacking in transparency. Some EAOs are concerned that dam development may progress further as a result of a ceasefire agreement, as development work has become easier to conduct in areas with peace agreements than in active conflict zones (Christophersen & Stave, 2018).

A symbolic hydropower plant is the Baluchaung No. 2 Hydropower Plant (also known as the Lawpita Hydropower Station in some areas), which was constructed in Kayah State as part of the first postwar compensation project conducted by Japan. After the Baluchaung No. 2 Hydropower Plant began operating in 1960, it was expanded to a total capacity of 168 MW and 1200 GWh of electricity generated annually, which makes up about 14% of total power generation in Myanmar in 2010 (Japan International Cooperation Agency, Nippon Koei Co. Ltd, & Tokyo Electric Power Company, 2012). Taken together with the Baluchaung No.1 Hydropower Plant, which provides 28 MW and began operation in 1992 the Baluchaung plants have undoubtedly fulfilled a vital role in the provision of electric power in Myanmar. In 2014 the IPP began operation of the Baluchaung No. 3 Hydropower Plant with 52 MW of capacity (Lwin, 2014). The Baluchaung No. 1, No. 2, and

No. 3 plants are cascade-type, allowing them to use hydropower resources efficiently (Japan International Cooperation Agency, Nippon Koei Co. Ltd., & Tokyo Electric Power Company, 2013).

However, in the 1960s when operation began under the military regime, the plant was located in Kayah state, where KNPP is active. Initially, the plant did not transmit power locally, but only to the cities of Yangon and Mandalay, therefore bringing no benefits to the area. (Since 2015 the plant has provided power to Loikaw in Kayah state and Mobye and Phekon in Shan state.) To build the plant, 1740 people were forcibly relocated from the site (Kramer, Russell, & Smith, 2018) and received no compensation for the seizure of their land. Also, local residents suffered damage to agriculture and fisheries resulting from the changes in the river basin and had to work as security forces for the transmission towers, suffering injuries from landmines buried around the power plants and transmission towers (Burke et al., 2017). Not only did some local residents die or become disabled as the result of stepping on landmines, but if one of their livestock were to step on a landmine and be injured, they were also forced to pay penalties to the Burmese military in compensation for damage to the mines, which were military property (Kubo, 2014; Pyi Pyi Thant, interview, February 18, 2019).

In recent years, progress has been made with respect to the provision of power within the state, and the electrification ratio has exceeded the national average at 77% (Du Pont, 2019). The provision of electric power to nearby villages has also become more stable. Wooden utility poles have been replaced by concrete ones, and blackouts caused by strong winds have become nearly nonexistent (members of a village development committee, interview, March 21, 2019). Blackouts are now limited almost exclusively to planned outages to conduct repairs (members of a village development committee, interview, March 21, 2019).

5.3.1.2 Roads and bridges

Road construction is a highly sensitive matter that can easily worsen a conflict situation. The military can access an area more easily with the new road, so suspicion against the construction of bases by Tatmadaw and the road's potential use for military logistics purposes with respect to EAO-controlled areas runs deep. There are also concerns over forced appropriation (without compensation) and forced emigration with the development of large-scale roads. In a new stretch of the Asia Highway that connects

Thailand and Myanmar and was built in 2015, a fight broke out between DKBA and Tatmadaw in which two villagers were casualties and over 1000 individuals were displaced (Downing, 2016; Karen Human Rights Group, THWEE Community Development Network, & Karen Environmental, & Social Action Network, 2016). In the KNU-controlled area, which was a signatory of NCA in 2018, 2300 people were displaced as a result of the road upgrade by Tatmadaw (South et al., 2018). This led to a greater sense of distrust on the KNU side and led to a temporary leave by the KNU (Anderson, 2018); the KNU even left the NCA temporarily in 2018 (Nyein, 2018a). Taking a conflict-sensitive approach and understanding local circumstances are very important in the construction of a road.

5.3.2 Trust building through cooperative projects

As was discussed earlier in the chapter, there is a lack of trust between the government and EAOs, both of which are important stakeholders in peace processes in Myanmar. It has also been shown that large-scale development does not simply lead to peace development. This raises the following question: What triggers progress, however incremental, in a peace process? There is no doubt that trust between stakeholders is essential (Johnson & Lidauer, 2014).

Burke et al. (2017) has summarized points in a development project that contribute to the peace process without entrenching the conflict:

1. Returns for the local people
2. Sufficient consultation with local stakeholders, including the conflicting organization
3. Planning in a bottom-up manner and leaving decision making to the locals or enhancement of local authority and development of capacities so that decision making can be trusted
4. Ensuring that the standard of living is improved
5. Ensuring that the scheme aligns with the framework for peace process[5]

[5]The NCA states in Chapter 6 Article 25 that signatory EAOs have responsibilities related to (1) health, education, and socioeconomic development; (2) environmental conservation; (3) efforts to preserve and promote ethnic culture, language, and literature; (4) peace and stability and maintenance of rule; (5) receiving aid from donors for regional development and capacity building; and (6) permitting the execution of a project related to eradication of illicit drugs ("The Nationwide Ceasefire Agreement Between the Government of the Republic of the Union of Myanmar and the Ethnic Armed Organisations," 2015; "The Nationwide Ceasefire Agreement Between the Government of the Republic of the Union of Myanmar and the Ethnic Armed Organisations," 2015).

6. Prevention of aggravation of points of dispute (e.g., language education for the ethnic minority)
7. Ensuring that one side is not gaining any military benefits (e.g., road construction to conflict areas)
8. Accepting a diverse system rather than pushing the government system forward, including respect for existing service providers and consideration of collaboration

Through aid, local people can actually experience the dividends of a peaceful relationship between the state, EAOs, or ethnic communities and may improve their collaboration at a grassroots level. Also, if the project is managed through cooperation among all sides, villagers would not need to fear to draw the anger of one side as a result of receiving benefits from the project of others (Jolliffe, 2014). Specific success cases include an initiative in the health sector, and efforts are also underway for education.

5.3.2.1 Health

Cooperation in the healthcare sector between the government and healthcare providers in rural or ethnic minority areas would improve the standards of care and lead to an increase in trust among stakeholders, which in turn would contribute to the development of peace (Tang & Zhao, 2017). Specifically, the department of the government cooperates with CSOs that are also fully supported by EAOs and local communities in EAO-controlled areas to provide vaccination services to residents with coordination assistance from the United Nations Children's Fund (United Nations, 2016).

5.3.2.2 Education

Education is an area the importance of which everyone can agree on, but some fear that efforts in this area would encourage conflict (Christophersen & Stave, 2018). In particular, education for ethnic minorities in their own language is a critical problem. The government has approved education in ethnic languages if needed in Article 43 of a 2015 amendment to the National Education Law (Union of the Republic of Myanmar, 2014) and to start from primary level in Article 44. Until this law, primary education was officially conducted entirely in Burmese. However, formal teaching staff in official schools continue to be predominantly Bamar in ethnicity. These teachers are sent to ethnic minority areas by the government. However, there is a chronic shortage of teaching staff in ethnic minority areas, and villagers have discretion to hire staff needed to fill the deficiency at the expense of the villagers, so they could hire people of same ethnic group. (Johnson (anonym) of an international CSO, personal communication, February 12, 2019). Since 2018, five of the seven state governments have worked on developing a curriculum to implement the teaching of an ethnic minority language in school (Salem-Gervais & Raynaud, 2019). Education in ethnic minority languages has been in high demand with EAOs, and discussion to coordinate government systems and existing ethnic minority schools has just begun (Hirschi, 2019).

5.3.3 Energy access projects

5.3.3.1 Distributed renewable energy sources

As discussed in other chapters, the presence of distributed power generation is rising as a result of the fall in price for solar panels and batteries in recent times. IRENA reports that 63% of the generated electricity that is needed to achieve universal electricity access by 2030 is being supplied by off-grid electrical sources [minigrids: 44%, stand-alone systems such as solar home system (SHSs): 19%] (IRENA, 2017).

5.3.3.1.1 Solar home systems

The SHS combines a solar panel, a battery, and a controller and has a battery charge port for mobile phones as well. The solar panel is installed on the roof of a residence for uses such as lighting by wiring the panel to the indoors. A particularly small system known as a solar lantern is used for lighting and consists of a charge port, a small solar panel, and a battery. More expensive models of the SHS include larger capacities for solar panels and batteries and enable the use of radios, small televisions, and DVD players.

5.3.3.1.2 Minigrid

A minigrid is a system in which small-scale power stations such as solar, small hydropower, and biomass are connected to one or several villages by a grid. In rural areas with sparse populations the extension of bulk transmission networks is often inefficient in terms of cost and time. A minigrid can be an effective solution in such cases. While it depends on the capacity of a power plant and its backup (battery and diesel generator), a minigrid allows the use of equipment that consumes more electricity than the SHS can, thus enabling more productive use.

5.3.3.1.3 Electrification by distributed power sources

Distributed power generation is an effective approach to electrification. Distributed power generation is expected to have fewer elements that could encourage conflict than the extension of bulk transmission network in ethnic minority areas.

In Myanmar the main power grid is under the jurisdiction of the government. Power generation that connects to the main grid is under the jurisdiction of the central Ministry of Energy and Electricity (MOEE). Transmission of over 33 kV is managed by the MOEE, while 11-kV distribution lines are managed by the state and region governments. The 400-V distribution lines are managed by the communities themselves (Du Pont, 2019). Urban residents who live in an area where a distribution line extends to the proximity of their homes are required to pay only a connection fee, while villagers need to bear the cost of the transformers as well (Aung Myint, interview, February 21, 2019). The cable fee for a single village to connect to a grid is $20,000/mile, while the cost of a medium-voltage cable through which several villages can connect to a grid is $35,000/mile; a transformer for a single village costs over $10,000, and the connection fee for a household is in the range of $150–800 (Langre, 2018). The current system of main grid extension enhances unfairness.

In a main grid extension, somewhat large-scale work cannot be avoided. As was noted earlier, large-scale construction work tends to be viewed with suspicion because it is perceived as seeking to expand the authority of the central government. Connection to a grid can benefit the local people directly, but it is still inevitably seen as a government-led project. Extension of the grid to an EAO-controlled area requires a more sensitive adjustment, and the project may become a long-term initiative. On the other hand, distributed power generation can be introduced independently of the central government. In an actual operation, however, particularly in cases in which foreign institutions are involved in implementing the project, adjustment and coordination with both the state government and the EAO is done in advance (Jones and Young (anonym) of an international CSO, personal communication, November 13, 2018).

Given that villages in ethnic minority areas are often located in mountainous areas and houses are sparsely settled, SHS may be more appropriate than the minigrid for initial electrification (Young (anonym) of an international CSO, personal communication, November 13, 2018). Compared to rural villages in Bamar that are located in the central dry zone, the ethnic minority villages are often located at high altitudes. This is in part because states where ethnic minorities live are found in peripheral mountainous areas and in part because the villages were burned down by the Tatmadaw if the villagers were seen to be cooperating with the EAO during a conflict (irrespective of whether the village was actually cooperating or not), and houses were more susceptible to being burned down if they were concentrated in terms of their location (Jackson (anonym) of an international CSO, personal communication, December 10, 2018). This is why houses are located in places that are more difficult to find among the mountains and are harder to access. Installing a distribution line that connects these homes would be expensive.

5.3.3.2 National programs related to energy access

Table 5.5 shows projects related to energy access in ethnic minority areas.

5.3.3.2.1 National electrification project

The NEP aims at universal access by 2030 by providing on both on-grid and off-grid electrification (World Bank, 2019). For off-grid efforts, a subsidy is provided for the implementation of minigrids, and SHS are being distributed. On-grid efforts are handled by the MOEE, while off-grid efforts are managed by the Department of Rural Development (DRD) of the Ministry of Agriculture, Livestock and Irrigation of Myanmar. As of September 2018, the implementation of minigrids was limited to Burmese areas (Rodriguez and Lewis (anonym) of a ministry, personal communication, September 13, 2018). SHS are limited to regions that are controlled by the government (Young and Jones (anonym) of an international CSO, personal communication, November 13, 2018).

5.3.3.2.2 The national community driven development project

The National Community Driven Development Project (NCDDP) aims to improve access to small-scale basic infrastructure and services for rural communities. The community itself decides what is implemented or rehabilitated and decides how the funds are used (National Community Driven Development Project, 2017). NCDDP states that it targets conflict-affected townships and considers villages that are not registered with government to be ideal sites for its projects (World Bank, 2016). However, of approximately 40,000 total subprojects in the six years leading up to FY 2018–19, transportation-related infrastructure such as roads, bridges, footpaths, and jetties accounted for over half of the projects, while electrification projects accounted for fewer than

Table 5.5 energy access projects in ethnic minority area.[6,7,8,9]

	Period	Total amount	Notes	Donor
NEP	2015–21	USD 527 million	Grid extension: USD 321.25 M Off-grid electrification: USD 172.00 M Technical assistance: USD 20.00 M Contingent emergency: 13.75 M response	World Bank
NCDDP	2012–21	USD 535.5 million	Transport: 53% Education: 14% Water/sanitation: 12% *Electrification: 9%* (by cumulative number of subprojects by the end of FY 2018/19)	World Bank, Italy
Electrification project in Kayin state and eastern Mon state	2016–19	USD 4.5 million (JPY 495 million)	SHS installation, technical assistance to operators and villagers	Japan
Smart Power Myanmar	2018-		The facility was established in 2018. It is managed by Pact, CSO.	The Rockefeller Foundation, World Bank, USAID, and Yoma Strategic Holdings
The barefoot project	2017–19	USD 400k		Denmark, India and Finland, as well as the Energy and Environment Partnership (EEP)
Renewable energy support in Kayin State	2018–19	USD 100k		Denmark, Sweden, European Climate Foundation

Source: in the text below.

[6]Grid extension: USD 321.25 M, Off-grid electrification: USD 172.00 M, Technical assistance: USD 20.00 M, Contingent emergency: 13.75 M response
[7]Total of Community Block Grants (USD 356.55 M), Facilitation and Capacity Development (USD100.00 M), Knowledge and Learning (USD 6.00 M), Implementation Support (USD 54.00 M), Emergency Contingency Response (USD 18.95 M) (Bradley, 2019).
[8]1 USD = 110.201 JPY (XE.com Inc., 2019)
[9]Total of four projects.

10% of total projects (National Community Driven Development Project, 2018b). Details such as whether the electrification project is on-grid or off-grid are not being shown (Bradley, 2019; National Community Driven Development Project, 2018a; 2018b). Table 5.6 shows the number of electrification subprojects in ethnic minority areas.

The NCDDP has a favorable reputation because of its community-oriented approach, while others are concerned with the quality of the infrastructure projects that have been implemented or rehabilitated because the number of engineers available within Myanmar is insufficient to address the sheer number of subprojects (Tsuji, interview, February 16, 2019).

5.3.3.2.3 Electrification project in southeastern Myanmar

An international NGO known as BHN Association installs SHS in parts of KNU-controlled areas in Kayin State and Mon State. The project is funded through the Grant Aid Project of the Japanese Government in partnership with the Nippon Foundation, and the selection of contractors is managed through a bidding process, arrangement, and monitoring after completion. To train people to become competent in SHS operation and maintenance in current and future target villages, volunteers are solicited from among the villagers and are educated in the basics of electricity and dispatched to do actual installation work (Ministry of Foreign Affairs of Japan, 2018).

Table 5.6 Number of electrification subprojects in ethnic minority areas.

	No. of electrification subprojects
Kayin state	2 (Kyerinseikgyi Township)
Chin state	1 (Tonzan Township)
	12 (Matupi Township)
Mon state	3 (Bilin Township)
	6 (Chaungzon Township)
	6 (Paung Township)
Rakhine state	4 (Ann Township)
Shan state	13 (Namhsan Township)
	5 (Mabein Township)
Tanintharyi region	2 (Tanintharyi Township)
Kachin state	0
Kayah state	0

Source: [Lee (anonym) of ministry, email, November 2, 2018].

5.3.3.2.4 Smart power Myanmar

Smart Power Myanmar was established by Pact and The Rockefeller Foundation aims to mobilize funds to support the roll-out of minigrids and other solutions through public-private partnerships for rural electrification. The founding members are the Rockefeller Foundation, the World Bank, the U.S. Agency for International Development, and Yoma Strategic Holdings (Pact, 2018).

5.3.3.2.5 The Barefoot project

The Barefoot Project provides SHS to rural households in Myanmar. It aims to empower women as well, as women in villages are trained to install and maintain the SHS (WWF, 2018). The project was set up to run for two years from 2017 to 2019, and its budget was around USD 400,000 (Manandhar, email, June 6, 2019).

5.3.3.2.6 Renewable energy support in Kayin state

WWF and the Kayin state government signed a memorandum of understanding for advisory and assistance on renewable energy planning (Myanmar Energy Monitor, 2018; Thit, 2018). The term was for one year during 2018–19, and the budget was around USD 100,000 (Manandhar, email, June 6, 2019).

5.4 Stakeholder perspectives

In our study, we conducted semistructured interviews.[10] A semistructured interview is an open framework that enables communication in both directions through conversations. Unlike questionnaires, detailed questions do not need to be prepared in advance if mutual associations between topics are clearly identified before conducting the interview. Questions can be added during the interview, and flexibility is afforded to both the interviewer and interviewee (Food and Agriculture Organization of the United Nations, 1990).

Given that the topic of this study involved sensitive matters such as peace processes and conflict, the authors listened to the interviewees through a more open interview or through a conversation when the semistructured form was difficult to conduct or the topics were too difficult for the interviewee to accept. Table 5.7 shows the list of interviewees. Open and semistructured interviews

[10]A different analysis of the identical interview results has been presented by Numata et al. (2021).

Table 5.7 The list of interviewees.

	Semistructured	Unofficial/open
Myanmar government official	1	3
Foreign government officials		2
International CSO	8	11
Local CSO	6	1
International organization	1	1
Academia	2	
Village committee member		6
Total	18	24

were conducted from September 10–14, 2018; December 3–13, 2018, February 10–23, 2019; and March 19–23, 2019. They were conducted in Yangon, Nay Pyi Taw, Hpa-an of capital of Kayin State, Ah Lel Chaung Village Tract, PhoungPyar Village, Lawksawk Township (Burmese: Yatsauk) Township, Taunggyi District, Shan State, and Ngwe Taung Village, Ngwe Taung Village Tract, Demoso Township, Loikaw District, and Kayah State. Additional interviews were conducted face-to-face and via Skype in Tokyo, Japan, during the same period. Kayin State and Southeast Myanmar were the prime areas of activity of the interviewees of the local CSOs. In Kayin State, the dominant EAO was KNU who was one of the leading EAOs among the NCA signatories. Therefore international CSOs have been more active in Kayin state, and we were able to conduct interviews with multiple stakeholders there. The list also includes interviewees who were in charge of the overall peace process in Myanmar. Table 5.8 lists semistructured interviewees.

5.5 Findings and conclusions

Energy access is the issue to be resolved in Myanmar. To achieve universal access, the peripheral areas with ethnic conflicts need to be electrified. There has been energy access injustice in the ethnic minority areas, although it has not been discussed widely enough. The issue is not only in the ethnic minority areas but also in the rural Bamar area where the majority population lives. A national program such as the NEP off-grid electrification can reach the government-controlled area, but it may still be difficult to access the EAO-controlled area. Ethnic minority CSOs have

Table 5.8 List of semistructured interviewees.

Name	Affiliation	Location	Date
Saw Kaw Muh Too (Henry)	BHN	Hpa-an, Kayin State, Myanmar	2/11/2019
Zar Ni	BHN	Hpa-an, Kayin State, Myanmar	2/11/2019
	BHN	Hpa-an, Kayin State, Myanmar	2/11/2019
	BHN	Hpa-an, Kayin State, Myanmar	2/11/2019
Nicola Williams	The Asia Foundation	Yangon, Myanmar	2/22/2019
Kyi Phyo	MEE Net Myanmar	Yangon, Myanmar	2/22/2019
Peter Barwick	United Nations	Yangon, Myanmar	2/21/2019
Naw Thet Thet Htun	Karen Women Empowerment Group	Yangon, Myanmar	2/21/2019
Aung Myint	Renewable Energy Association Myanmar (REAM)	Yangon, Myanmar	2/21/2019
Pyi Pyi Thant	Heinrich Böll Stiftung Myanmar	Yangon, Myanmar	2/18/2019
Khin Ma Ma Myo	Myanmar Institute of Gender Studies	Yangon, Myanmar	22/17/019
Fukio Tsuji	Peace Winds Japan	Hpa-an, Kayin State, Myanmar	2/16/2019
Saw Tha Moo	Local organization	Yangon, Myanmar	2./15/2019
Ugan Manandhar	WWF Myanmar	Yangon, Myanmar	2/13/2019
Akio Takahashi	Institute of Advanced Studies on Asia, The University of Tokyo	Tokyo, Japan	3/11/2019
Ashley South	Independent researcher	Skype	3/27/2019

worked to improve the villagers' livelihood in these areas. Some local CSOs have provided social services, but energy and electricity have often been excluded.

Regrettably, energy has often been associated with resource exploitation because of past experiences with large-scale hydropower development in conflict contexts. However, technological innovations now make off-grid renewable power sources more affordable. SHS is easy to deploy and is not politically divisive. Its use is mostly limited to lighting and mobile phone charging. However, it can make a major change in people's means of livelihood. Mobile phones and social networking services have spread rapidly all over the world, and rural areas in Myanmar are no exception. SHS is considered to suit local needs.

The peace process in Myanmar has reached a stalemate, and trust among stakeholders has not been built. The current framework seems to take a top-down approach, such as the Union Peace Conference and large-scale economic development. A bottom-up approach is necessary to move the situation forward. Both sides, the government/Tatmadaw and EAOs, can agree that some improvement in the standard of living among villagers in conflict areas is a matter of great urgency. Projects to support livelihoods do not only provide returns to local villagers but also create opportunities for trust building at the grassroots level. Collaboration among state and region governments, EAOs and ethnic minority groups, and CSOs is necessary to ensure the success of energy sector projects. In other words, all stakeholders must meet often and remain in constant contact. Frequent meetings are important in building relationships in Myanmar. Collaboration in the health sector is ongoing. Off-grid renewables can be another option. Overall, it is better to have multiple connections among stakeholders in different sectors.

The following are our findings obtained from the interviews.

5.5.1 It is important that aid meets the needs of a community

Aid programs must understand local needs (South, interview (Skype), March 27, 2019; Williams, interview, February 22, 2019). For this to be achieved, local stakeholders need to be engaged, and the needs of the community must be understood thoroughly. Otherwise, a minor malfunction may be enough for the program to be abandoned.

5.5.2 Trust can be built through bottom-up cooperative projects

Collaborations between the state government and EAOs or the ethnic minority CSOs are effective in building trust-based relationships from the bottom up. Building up cooperation over "low-hanging fruit," or areas in which the stakeholders can easily agree, will be effective in fostering a trusting relationship from the bottom up. Initiatives are already underway in areas of health and education, and distributed energy can be another example of such initiatives, but current efforts with respect to distributed energy are limited.

5.5.3 Frequent exchange and contact are important

Cooperative projects create opportunities to meet stakeholders. In Myanmar, frequent face-to-face meetings between parties are important in building and sustaining relationships. Relationships based on personal networks are vital for relationships among organizations. In addition to the Union Peace Conference–21st Century Panglong, which is a large conference that occasionally takes place at a top level, bilateral meetings that take place frequently at the grassroots level are meaningful (Takahashi, interview, March 11, 2019).

5.5.4 Further diffusion of mobile phones assisted by solar home system may be able to narrow the digital divide

There is a strong need for mobile phones, which in turn need electricity for recharging the batteries. In response to the question of how widely spread mobile phone use is in rural areas, the answers varied based on the interviewees. Some said that almost everyone had mobile phones, even in rural areas (Takahashi, interview, March 11, 2019; Tsuji, interview, 16, February 2019; Aung Myint, interview, February 21, 2019; Anderson (anonym) of a local CSO, personal communication, December 9, 2018). Other respondents pointed out that many people in ethnic minority villages still did not own mobile phones, while others had connectivity issues with mobile networks in the vicinity (Allen (anonym) of local CSO, interview (Skype), February 28, 2019; Smith (anonym) of an international CSO, personal communication, February 12, 2019; Young (anonym) of an international CSO, personal communication, November 13, 2018). In 2017 the overall mobile phone penetration rate had reached 81.5% in Myanmar, with an average of 76.6% penetration in the rural areas of the country. An analysis showed that the gap in ownership rates between urban and rural areas was likely because of the difference in purchasing power rather than because of the underdevelopment of the infrastructure (Central Statistical Organization, UNDP, & World, 2018). There are cases in which a powerful figure in the village has given away a used mobile phone to a villager (Takahashi, interview, March 11, 2019). Throughout the interviews, the need for mobile phones were pointed out very strongly, especially among the younger people (Aung Myint, interview, February 21, 2019; Anderson (anonym) of a local CSO, personal communication, December 9, 2018). Charging mobile phones and lighting can be covered by the use of solar lanterns. If mobile phone batteries can be charged by using distributed generation of power, villagers will no longer need to pay to charge their phones at shops with relatively expensive prices. In line with the global trend, social networks have spread tremendously in Myanmar. Connecting to a social network is not only strongly desired but is also believed to contribute to the reduction of the digital divide in rural areas.

5.5.5 Many solar home system have too short a duration, implying the need to improve the quality of solar home system in the market

Some interviewees thought that the lifespan of a SHS is approximately two years (Tsuji, interview, February 16, 2019; Aung Myint, interview, February 21, 2019; Williams, interview, February 22, 2019). The literature also reports that the battery for SHS lasts 1.5–2 years (Manhar, Latt, & Hilbert, 2018). In analyzing aspects of engineering, it might appear that the battery life is much shorter than that of other options, since many believe that the battery lasts for 25 years or more for a solar panel, 10 years for a lithium battery, and 4–5 years for lead. However, two years of life is the realistic operating life for products that are available in the market. Since they are available at an affordable price, many interviewees tend to check whether the light turns on or not at the time of purchase and have the attitude that they can simply buy a replacement if the product breaks (Aung Myint, interview, February 21, 2019).

5.5.6 Roads, which are a prerequisite for development, must be constructed in consultation with local communities, reflecting the local contexts

Many interviewees pointed out how crucial roads are to the development of a village (Zar Ni, interview, February 11, 2019; Tsuji, interview, February 16, 2019; Takahashi, interview, March 11, 2019; Anderson (anonym) of a local CSO, personal communication, December 9, 2018). When roads are built and people move to and from villages, products from the village can be sold in markets, and an industry can be created. Social services can also be accessed more easily. Some of the villages in the ethnic minority areas become isolated during the rainy seasons, and the importance of access is recognized widely. Many communities welcome the construction and upgrades of roads and bridges (South et al., 2018). However, roads can disrupt a fragile situation. It is important to consider the local context and to facilitate the appropriate consultation process.

Appendix: a brief history of ethnic armed organizations

The British colonial period

Two methods of governance were used during the British colonial period. One was direct governance, which was used in the flat plain regions of present-day Tanintharyi region, Rakhine state, Bago region, and Ayeyarwady region, which were then called Burma proper or ministerial Burma. In contrast, peripheral areas such as the Shan kingdom were left autonomous (Burke et al., 2017; Myint-U, 2011; Nemoto, 2014). The colonial government also supported the construction of Christian mission schools, which then expanded mainly among the Kachin, Chin, and Karen peoples (Encyclopedia Britannica, 2018). The colonial government appointed members of minorities who were primarily Christian to be part of the military and the police, thereby connecting them to British rule (Myint-U, 2011).

At that time, Myanmar was a part of the British colony of India, and many people of different classes from India, including merchants and moneylenders, began to migrate into Myanmar. Cases often occurred in which Bamar farmers gave up their land to repay debts owed to Indian moneylenders (Encyclopedia Britannica, 2018). In colonial period Burma, the upper class was made up mainly of the British; the middle class was made up of Chinese, Indians, Karen, and Bamar; and the lower class was made up of working-class Indians and Bamar (Nemoto, 2014). Antagonism between ethnic groups deepened because of this situation. However, it is not sufficient to simply say that a firm ethnic consciousness formed during the colonial period and led to the outbreak of civil war, as misgovernment following independence also contributed significantly (Kubo, 2014).

Kayin state[11]

Independence from the British following the Second World War was also the beginning of the long period of conflict between ethnic groups. During the Second World War, the independence forces of General Aung San and others switched from cooperating with the Japanese to resisting them, and after the war the resistance turned to wresting independence from the British. During this series of political events, the first Panglong

[11]Kayin, the name of the state as established by the Myanmar government, has been used, but Karen is used as the name of the ethnic group.

Agreement was concluded in February 1947 and signed by General Aung San and various minority peoples. However, the minority peoples who participated were limited to the Shan, Kachin, and Chin; the Karen and Karenni (or Kayah) were mere observers. Also, no members of the Mon and Arakan peoples attended the conference. Aung San was assassinated shortly after in July 1947. His successor, U Nu, became prime minister, and Myanmar gained independence from Britain in 1948. However, because the priority had been given to independence itself, the Karen State that the Karen people demanded was not established at the outset, and conflicts with the KNU, which was established in 1947, began in 1949 (Nemoto, 2014). The national military has used methods such as encouraging the internal collapse of EAOs by approaching dissatisfied lower-ranking members of the internal EAO leadership (Jolliffe, 2014). As a result, EAOs that have separated or rejoined the KNU during the long history of the conflict include the NCA signatories of the DKBA and the KNU/KNLA-PC (Saferworld, & Karen Peace Support Network, 2019).

The 8888 uprising

The democratic movement that was inspired by student fatalities in collisions between students at the Rangoon Institute of Technology (today Yangon Technological University) and police in 1988 brought about the retirement of Ne Win but was also connected to a Tatmadaw coup d'état and ultimately the beginning of the military regime that lasted until 2011. The students who had formed the democratic movement organized the ABSDF and became active in areas controlled by the KIO and KNU (Myanmar Peace Monitor, 2016).

The military regime, which feared that the student movement would strengthen the movement for democracy, moved to clamp down on the universities, closing undergraduate education at Yangon Technological University until 2011 and at Yangon University until 2013 with only the graduate school accepting students (University Of Yangon, n.d.; Yangon Technological University, n.d.). This caused a shortage of well-educated human resources in Myanmar.

The push for a bilateral ceasefire

The push for a bilateral ceasefire began to expand in the 1990s. The Thai government, which had deepened ties with the Myanmar government since the second half of the 1980s, placed pressure on

EAOs in areas along the Thai border to participate in peace negotiations. The Thai government also pressured a group of 8000–10,000 IDP of the Mon ethnic group to return to Myanmar (Jolliffe & South, 2014). In 1995 the NMSP, which was primarily active in Mon state, concluded a ceasefire agreement with the government (Kramer, 2012). Although ceasefire agreements had been concluded with a number of groups, the push for ceasefires waned with the 2004 overthrow of the Khin Nyunt, who had supported their implementation (Durieux & Dhanapala, 2008). Peace was left unachieved, and the next push for ceasefires would have to wait for the beginning of the Thein Sane administration.

Nationwide Ceasefire Agreement nonsignatories

Among NCA nonsignatories are the allied groups of the Northern Alliance, which have been active and have engaged in many battles in recent years (Raleigh, 2018). The Northern Alliance was established in 2016 by four non-ceasefire EAOs: the KIO, AA, TNLA, and MNDAA. The KIO spearheaded the establishment of the Northern Alliance (Bynum, 2018). After this, these four EAOs established the Federal Political Negotiation and Consultative Committee together with three other EAOs: the UWSA, SSPP, and NDAA (Tønnesson et al., 2019). As will be discussed later in this appendix, the UWSA boasts the largest forces of any EAO in Myanmar (Yun, 2017). The UWSA is also thought to maintain a close relationship with China, which touches the border of that special administrative district (SAD) (Myint-U, 2011), and has had an impact on the four EAOs of the Northern Alliance (Tønnesson et al., 2019). China is also thought to be acting as a broker between the Northern Alliance and the Myanmar government (Mangshang & South, 2019). The China-Myanmar Economic Corridor (CMEC), which is proceeding under the Chinese "One Belt, One Road" initiative, runs from the town of Muse in Shan state through the former capital of Mandalay and connects to Kyaukphyu in Rakhine state. The condition of this route is of serious importance to China (United States Institute of Peace, 2018). Below, the origins of the main nonsignatories to the NCA among the main EAOs will be examined.

Ethnic armed organizations originating in the communist party of Burma

After independence from Britain, one of the combatants that the independent government had to fight was the Communist Party of Burma (CPB). This CPB had been established in 1939, primarily by Aung San, but it fractured after the Second World War and entered into armed conflict with the government following independence (Bi, 2012). In addition to the civil war, the Kuomintang invaded Shan state around 1950, having lost its war against the Chinese Communist Party (Nemoto, 2014). In the early 1960s the China-Myanmar relationship was positive enough for the government to cooperate with China in mopping up the army of the Kuomintang. However, in 1962 the military regime began, following the coup d'état by Ne Win, and China shifted its own policies to support foreign communist parties. This changed the Myanmar-China relationship, which deteriorated further as a result of the revolt in 1967. As the relationship between the two countries worsened, the Chinese government turned to supporting the CPB. The CPB controlled 100,000 km^2 at the height of its power, including nearly all of the border between China and Myanmar (excluding Muse) (Bi, 2012; Kramer, 2012; Kramer et al. 2018). However, beginning in the latter half of the 1970s, the Chinese government reduced its support for the CPB as a result of a shift in its foreign policy. Although the CPB proceeded to produce opium as a separate source of funds, it ultimately collapsed internally in 1989, with each ethnic group forming its own EAO: in northeast Shan State the MNDAA of the Kokang, the UWSA of the Wa, and the NDAA of the Mong La Shan and the Aka and in northeast Kachin state the New Democratic Army-Kachin (NDA-K) of the Kachin people. The military regime took this as an opportunity and began peace negotiations directed by Khin Nyunt, who would later become prime minister, and a ceasefire agreement was eventually concluded (Bi, 2012).

These four EAOs were allowed to turn the areas they controlled into SADs, where they engaged in sophisticated forms of self-rule (Bi, 2012). In these areas the narcotics business produced great wealth, and the UWSA became the world's largest producer of heroin (Myint-U, 2011). The cultivation of alternative crops was promoted under government leadership, but the cultivation of heroin was never eradicated (Bi, 2012).

Afterwards, the four EAOs took different paths. The UWSA obtained great wealth from the narcotics business and has more than 30,000 soldiers as well as weapons including surface-to-air missiles (Myint-U, 2011), is considered the "world's mightiest non-state army," and boasts the largest forces of any EAO (Yun, 2017). Most of the UWSA-controlled area is connected to the Chinese electric grid, Chinese is spoken, and the towns are prosperous (Myint-U, 2011). In the NDAA-controlled area, Mong La has become famous for casinos and the illegal wildlife trade (Myint-U, 2011). In contrast, the MNDAA broke up following armed conflict with the national military in 2009, and a portion of it has converted into the Border Guard Force, described later. In the same year the NDA-KI also converted the Border Guard Force (Kramer, 2012).

Kachin state and Shan state

In recent years, one of the areas where combat has most repeatedly broken out has been northeastern Myanmar, specifically in Shan state and Kachin state (Raleigh, 2018). Shan state has historically been governed by local lords (sawbwa), and even during the British colonial period, with the exception of those who opposed colonial rule, these lords were allowed to remain and their authority was reinforced (Myint-U, 2011). Those who converted to Christianity, such as the Kachin mountain people, were given appointments within the colonial government. Christian school were constructed in Kachin, which increased the level of education (Burke et al., 2017). At the beginning of Burmese independence, the Shan people agreed to participate in the Burmese government and federation. Their right to a certain degree of self-government was recognized, but owing to the effects of the civil war with the CPB, the situation in Shan state deteriorated. The local lords were arrested during the 1962 coup d'état. In comparison to the U Nu administration, the rights of minority peoples were weakened during the period of Burmese socialism, including the abolition of self-government (Nemoto, 2014). As the scope of the war expanded, the KIO was formed by the Kachin people in Kachin, and the Shan State Army[12] was formed in Shan state.

The military regime adopted the infamous "four cuts" strategy in the 1960s, which involved cutting off coordination between EAOs and villages in terms of food, funds, intelligence, and recruits. Under this strategy, villagers were unable to farm, and some were forcibly relocated to locations with no access to food or medical care (Burke et al., 2017; Smith, 1994). These forced relocations continued even after ceasefire agreements had been concluded with the leaders of the minority peoples (Japan International Cooperation Agency, Nippon Koei Co. Ltd., & Tokyo Electric Power Company, 2013; Jolliffe, 2014; Kubo, 2014).

[12]This became the base of the armed wings of the SSPP and RCSS, the Shan State Army-North and Shan State Army-South.

As a result, many villagers became IDP. Also, villagers whose villages were burned or who evacuated because of nearby combat also became IDP. The "four cuts" strategy is not merely a relic of the past; it is reported to have been used in conflicts between the AA of Rakhine state and the Tatmadaw in 2019 (Pwint, 2019). Also, 99,000 IDP were created in the three years of conflict that reignited in the northern part of the country beginning in 2011, such as in Kachin and Shan state (Benson & Jaquet, 2014; Visser, 2016). Repatriating these people and improving their living environment are major challenges (Arraiza & Leckie, 2018).

The KIO agreed to a ceasefire that lasted from 1994 until 2011 (Bynum, 2018). This ceasefire involved selling the abundant resources of Kachin state to Chinese companies, helped to line the pockets of both the Tatmadaw and the KIO leaders, and was unpopular with local residents (Tønnesson & Nilsen, 2018), However, there were certainly few conflicts in Kachin state in 2000, and the situation had become more stable. Both KIO and Tatmadaw soldiers were seen at local festivals in Kachin. Even in terms of observed IDP, none were created in Kachin in 2006, although IDP did occur in Kachin in 2016 (Burke et al., 2017). The impetus for the violation of the ceasefire is said to be the administration's demand that the KIO convert into a Border Guard Force (BGF) (Visser, 2016; Myanmar Peace Monitor, n.d.-a). In 2009 the administration demanded that EAOs be converted into BGFs based on the 2008 constitution and be put under the command of the Tatmadaw. The demand to convert into BGFs was in actuality a demand to but put under the control of the national armed forces, a difficult requirement for the EAOs to accept (Myint-U, 2011). In recent years the KIO has lost a lot of territory in combat and is thought to be leaning toward dialog with the government (Tønnesson et al., 2019).

Rakhine state

In recent years, Rakhine state has gained attention on account of the Rohingya refugee problem,[13] but the AA has also become increasingly active and engaged in more combat. The AA was organized comparatively recently, in 2009, and for a while was active in Kachin state, which is controlled by the KIO, with which the AA had a cooperative relationship. However,

[13]As was stated previously, the Rohingya refugee problem is extremely politically sensitive and is not dealt with in this chapter, as it has a different background and should be understood in a different way than the discussion of minority peoples contained herein.

from 2015 onwards, the AA moved its activities to Rakhine state and the southern part of Chin state (Tønnesson et al., 2019). At first the AA was thought to have had only 1000 members in 2011 (Burma News International, 2017), but it rapidly increased in size by recruiting through social network services and is now thought to have expanded to around 7000 members (The Irrawaddy, 2019). The AAs adversary is not the Rohingya but the Tatmadaw. The Tatmadaw appeared to be occupied with the Northern Alliance and in December 2018 announced a four-month unilateral ceasefire that covered Shan state and Kachin state in the northeast part of the country (Nyein, 2018b). Directly after this, in January 2019, the AA attacked four border guard police outposts in Rakhine state. Combat has since intensified (The Irrawaddy, 2019). (The four-month unilateral ceasefire did not include Rakhine state (Lat, Tun, & Thu, 2018)).

References

ADB. (2016). Myanmar: Energy sector assessment, strategy, and road map. Manila, Philippines. <https://www.adb.org/sites/default/files/institutional-document/218286/mya-energy-sector-assessment.pdf>.

Anderson, B. (2018). *Stalemate and suspicion: An appraisal of the Myanmar peace process.* <https://teacircleoxford.com/2018/06/06/stalemate-and-suspicion-an-appraisal-of-the-myanmar-peace-process/> Accessed 19.12. 18.

Arraiza, J., & Leckie, S. (2018). A vision for restitution in Myanmar. *Forced Migration Review, 57*, 72–74. Available from http://libproxy.lib.unc.edu/login?url = https://search.proquest.com/docview/2016974148?accountid = 14244%0Ahttp://vb3lk7eb4t.search.serialssolutions.com/?genre = article&atitle = A + vision + for + restitution + in + Myanmar&author = Arraiza%2C + José%3BLeckie%2C + Scott&volume = .

Asianews.it. *Rebel armies in peace talks in Shan state.* (2019). <http://asianews.it/news-en/Rebel-armies-in-peace-talks-in-Shan-State-46876.html> Accessed 10.05.19.

Aung, K.H. (2019). *Aung San statue controversy highlights vulnerability of ethnic minority identity.* <https://frontiermyanmar.net/en/aung-san-statue-controversy-highlights-vulnerability-of-ethnic-minority-identity> Accessed 26.03.19.

Aung, S.Y. (2018). *Still no date for release of census findings on ethnic populations.* <https://www.irrawaddy.com/news/burma/still-no-date-release-census-findings-ethnic-populations.html> Accessed 19.12.18.

Batcheler, R. (2018). State and region governments in Myanmar. Yangon. <https://asiafoundation.org/wp-content/uploads/2018/10/State-and-Region-Governments-in-Myanmar_New-Edition-2018_Eng.pdf>.

Benson, E., & Jaquet, C. (2014). *Faith-based humanitarianism in northern Myanmar.* <http://www.fmreview.org/en/faith/benson-jaquet.pdf> Accessed 01.11.18.

Bi, S. (2012). Kokkyo chiiki no shosu minzoku seiryoku wo meguru Chugoku Myanmar kankei (in Japanese). In T. Kudo (Ed.), *Myanmar seiji no jitsuzo:*

gunsei 23nen no kozai to shinseiken no yukue (pp. 167–199). Tokyo: Institute of Developing Economies, Japan External Trade Organization.

Billen, D., & Bianchi, G. Decentralised energy market assessment in Myanmar. Yangon. (2019). <https://www.pactworld.org/library/decentralised-energy-market-assessment-myanmar>.

Bradley, S. Disclosable version of the ISR – Myanmar national community driven development project – P132500 – Sequence No: 14 (English). Washington, DC (2019). <http://documents.worldbank.org/curated/en/236551547999761770/pdf/Disclosable-Version-of-the-ISR-Myanmar-National-Community-Driven-Development-Project-P132500-Sequence-No-14.pdf>.

Burke, A., Williams, N., Barron, P., Jolliffe, K., & Carr, T. (2017). *The contested areas of Myanmar: Subnational conflict, aid, and development.* <https://asiafoundation.org/wp-content/uploads/2017/10/ContestedAreasMyanmarReport.pdf>.

Burma News International. Deciphering Myanmar's peace process: A reference guide 2016. Chiang Mai. (2017). <https://www.bnionline.net/sites/bnionline.net/files/publication_docs/dm_peace_process_a_reference_guide_2016.pdf>.

Bynum, E. (2018). *Analysis of the FPNCC / northern alliance and Myanmar conflict dynamics.* <https://www.acleddata.com/2018/07/21/analysis-of-the-fpncc-northern-alliance-and-myanmar-conflict-dynamics/> Accessed 04.04.19.

Bynum, E. (2019). *Ceasefires and conflict dynamics in Myanmar.* <https://www.acleddata.com/2019/05/13/ceasefires-and-conflict-dynamics-in-myanmar/> Accessed 14.05.19.

Central Intelligence Agency. (2019). *The world fact book: East Asia/Southeast Asia::Burma.* <https://www.cia.gov/library/publications/the-world-factbook/geos/bm.html> Accessed 22.04.19.

Central Statistical Organization, UNDP, & World, Bank. (2018). Myanmar living conditions survey 2017: Key indicators report. Nay Pyi Taw and Yangon, Myanmar. <http://documents.worldbank.org/curated/en/739461530021973802/pdf/127618-WP-P162753-PUBLIC-MyanmarLivingConditionsLowRes.pdf>.

Christophersen, M., & Stave, S.E. Advancing sustainable development between conflict and peace in Myanmar. New York. (2018). <https://www.ipinst.org/wp-content/uploads/2018/04/IPI-Rpt-Myanmar.pdf>.

Department of Population Ministry of Immigration and Population. The 2014 Myanmar population and housing census the union report: Religion census report volume 2-C. Union of Myanmar (Vol. 2–C). Nay Pyi Taw. (2015). <https://doi.org/10.1080/09578810410001688815>.

Downing, J. (2016). *The old road through the mountains.* <https://frontiermyanmar.net/en/the-old-road-through-the-mountains> Accessed 17.05.19.

Durieux, J., & Dhanapala, S. (2008). Carving out humanitarian space. Forced Migration Review, 13–15. <https://www.fmreview.org/sites/fmr/files/FMRdownloads/en/burma/durieux-dhanapala.pdf>.

Embassy of the Union of Myanmar Brussels. *Composition of the different ethnic groups under the 8 major national ethnic races in Myanmar.* (n.d.). <http://www.embassyofmyanmar.be/ABOUT/ethnicgroups.htm> Accessed 20.12.18.

Encyclopedia Britannica. (2018). Myanmar. In Encyclopedia britannica. Encyclopaedia Britannica, Inc. <https://www.britannica.com/place/Myanmar>.

Ethnologue. *Myanmar languages.* (n.d.). <https://www.ethnologue.com/country/MM/languages> Accessed 20.12.18.

Everson, M., & Hosken, M. (2006). *Proposal for encoding Myanmar characters for Karen and Kayah in the UCS.* <https://www.unicode.org/L2/L2006/06303-n3142-myanmar-karen-kayah.pdf> Accessed 26.06.19.

Food and Agriculture Organization of the United Nations. (1990). Section three: The tools Chapter eight: The tools and how to use them Tool 9: Semi-structured interviews. In The community's toolbox: The idea, methods and tools for participatory assessment, monitoring and evaluation in community forestry. Rome: FAO. <http://www.fao.org/3/x5307e/x5307e08.htm>.

Hirschi, E. (2019). *Mon lead the way in mother tongue education.* <https://frontiermyanmar.net/en/mon-lead-the-way-in-mother-tongue-education> Accessed 29.06.19.

IFC. (2017). *Hydropower database.* Washignton, DC: IFC. Available from https://www.ifc.org/wps/wcm/connect/07b1eba5-9c63-4f7c-8674-d463732a0395/Copy + of + 18 + 02 + 07 + MYA + SEA + HP + Database + WB-IFC.xlsx?MOD = AJPERES.

Institute for Security & Development Policy. (2015). *Myanmar's nationwide ceasefire agreement.*

International Finance Corporation. (2018). Strategic environmental assessment of the Myanmar hydropower sector final report. International Finance Corporation. Washignton, DC. <https://www.ifc.org/wps/wcm/connect/2f7c35f4-e509-48b2-9fd8-b7cbc0501171/SEA_Final_Report_English_web.pdf?MOD = AJPERES>.

IRENA. (2017). REthinking energy 2017: Accelerating the global energy transformation (Vol. 55). Abu Dhabi.

Japan International Cooperation Agency, Nippon Koei Co.Ltd, & Tokyo Electric Power Company. (2012). Preparatory survey report on the project for rehabilitation for Baluchaung No. 2 Hydro Power Plant in the Union of Myanmar (1) (Baluchaung No.2 Suiryoku hatsudensho hoshuu keikaku junbi chosa (1) chosa hokokusho in Japanese). Tokyo. <http://open_jicareport.jica.go.jp/pdf/12088589.pdf>.

Japan International Cooperation Agency, Nippon Koei Co. Ltd., & Tokyo Electric Power Company. (2013). Preparatory survey report on the project for rehabilitation for Baluchaung No. 2 Hydro Power Plant in the Union of Myanmar (2) (Baluchaung No.2 Suiryoku hatsudensho hoshuu keikaku jizen chosa (2) in Japanese). Tokyo. <https://libopac.jica.go.jp/images/report/12114708.pdf>.

Johnson, C., & Lidauer, M. (2014). Testing ceasefires, building trust: Myanmar peace support initiative operational review. Oslo. <https://www.regjeringen.no/contentassets/3e66a5bcb64e4a96a950e2bcafb2d885/mpsi_report.pdf>.

Jolliffe, K. (2014). Ethnic conflict and social services in Myanmar's contested regions. The Asia Foundation. Yangon. <http://asiafoundation.org/publication/ethnic-conflict-and-social-services-in-myanmars-contested-regions/>.

Jolliffe, K., Bainbridge, J., & Campbell, I. (2017). Security integration in Myanmar: Past experiences and future visions. Saferworld. London. <https://www.saferworld.org.uk/resources/publications/1132-security-integration-in-myanmar-past-experiences-and-future-visions>.

Jolliffe, K., & South, A. (2014). New issues in refugee research (Research Paper No. 271). Geneva. <https://www.unhcr.org/533927c39.pdf>.

Karen Human Rights Group, THWEE Community Development Network, & Karen Environmental and Social Action Network. (2016). *Beautiful words, ugly actions: The Asian highway in Karen state.* <http://khrg.org/sites/default/files/beautiful_words_ugly_actions_-_english_for_web.pdf> Accessed 17.05.19.

Kean, T. (2019). *China advising on hydro policy as govt backs away from IFC assessment.* <https://frontiermyanmar.net/en/china-advising-on-hydro-policy-as-govt-backs-away-from-ifc-assessment> Accessed 13.05.19.

Kittner, N., & Yamaguchi, K. (2017). Hydropower threatens peace in Myanmar – But it doesn't have to. Nikkei Asian Review, March. <https://asia.nikkei.com/Economy/Hydropower-threatens-peace-in-Myanmar-but-it-doesn-t-have-to>.

Kramer, T., Russell, O., & Smith, M. (2018). From war to peace in Kayah (Karenni) State: A land at the crossroads in Myanmar. Amsterdam. <https://www.tni.org/files/publication-downloads/tni-2018_karenni_eng_web_def.pdf>.

Kramer, T. (2012). Ethnic conflict in Myanmar: Challenges for the new government (in Japanese). In T. Kudo (Ed.), *Myanmar seiji no jitsuzo : gunsei 23nen no kozai to shinseiken no yukue* (pp. 139–166). Tokyo: Institute of Developing Economies, Japan External Trade Organization.

Kubo, T. (2014). *Nanmin no Jinruigaku-Thai Burma kokkyo no Karennni nanmin no idou to teiju (in Japanese).* Tokyo: Shimizukobundo.

Kudo, T. (2010). Myanmar sosenkyo to sonogo (in Japanese). Asia No Dekigoto, 10, 1–7. <https://ir.ide.go.jp/?action = pages_view_main&active_action = repository_view_main_item_detail&item_id = 49588&item_no = 1&page_id = 39&block_id = 158>.

Lall, M., & South, A. (2018). Power dynamics of language and education policy in Myanmar's contested transition. *Comparative Education Review, 62*(4), 482–502. Available from https://doi.org/10.1086/699655.

Langre, G. de. (2018). The real cost of Myanmar's electricity. <https://www.mmtimes.com/news/real-cost-myanmars-electricity.html> Accessed 07.05.19.

Lat, W.K. K., Tun, W.M., & Thu, K. (2018). *Myanmar military declares four-month cease-fire in Shan, Kachin conflict zones.* <https://www.rfa.org/english/news/myanmar/army-ceasefire-12212018161704.html> Accessed 25.12.18.

Lwin, H. T. (2014). *Powering the society – Balu Chaung (3).* Singapore: Aspire 360 Pte Ltd.

Lynn, K.Y. (2019). *Interview, terminated: A bridge too far for Mon State Chief Minister.* <https://frontiermyanmar.net/en/interview-terminated-a-bridge-too-far-for-mon-state-chief-minister> Accessed 26.03.19.

Mangshang, Y.B., & South, A. (2019). *China, India and Myitsone: The power game to come.* <https://frontiermyanmar.net/en/china-india-and-myitsone-the-power-game-to-come> Accessed 04.03.19.

Manhar, A., Latt, K., & Hilbert, I. (2018). Report on the fact finding mission on the management and recycling of end-of-life batteries used in solar home systems in Myanmar. Freiburg & Yangon. <https://www.oeko.de/fileadmin/oekodoc/Batteries-from-SHS-Myanmar.pdf>.

Ministry of Electricity and Energy Electricity Supply Enterprise The Republic of the Union of Myanmar. (2019). *NEP plan.* <http://www.moee.gov.mm/en/ignite/page/80> Accessed 17.05.19.

Ministry of Foreign Affairs of Japan. (2017). *Republic of the Union of Myanmar Kiso data* (in Japanese). <https://www.mofa.go.jp/mofaj/area/myanmar/data.html> Accessed 08.05.18.

Ministry of Foreign Affairs of Japan. (2018). *Project completion report: Electrification project in central Western Kayin State and Eastern Mon State* (in Japanese). <https://www.mofa.go.jp/mofaj/files/000426157.pdf> Accessed 16.05.19.

Myanmar Energy Monitor. (2018). *WWF signs MoU for renewable energy support in Kayin State.* <https://energy.frontiermyanmar.com/news/renewables/wwf-signs-mou-renewable-energy-support-kayin-state> Accessed 28.05.19.

Myanmar Peace Monitor. (2016). Armed ethnic groups 2016. Yangon: Myanmar Peace Monitor, The Burma News International. <http://www.mmpeacemonitor.org/#!armed-ethnic-groups/c1q70>.

Myanmar Peace Monitor. *Kachin independence organization.* (n.d.-a). <http://www.mmpeacemonitor.org/stakeholders/myanmar-peace-center/155-kio> Accessed 25.04.19.

Myanmar Peace Monitor. (n.d.-b). *Stakeholders overview.* <http://mmpeacemonitor.org/stakeholders/stakeholders-overview> Accessed 25.04.19.

Myint-U, T. (2011). *Where China meets India: Burma and the new crossroads of Asia.* New York: Farrar, Straus and Giroux.

Naing, S.Y. (2017). *Behind the KNU election results.* <https://www.irrawaddy.com/opinion/commentary/behind-the-knu-election-results.html> Accessed 16.05.19.

National Community Driven Development Project. (2017). *Project annual report (April 2016- March 2017) executive summary.* <http://cdd.drdmyanmar.org/sites/cdd.drdmyanmar.org/files/documents/final_annual_report_exe_summary_aprial_2016-_march_2017.pdf> Accessed 02.11.18.

National Community Driven Development Project. *NCDDP community driven development project.* (2018a). <https://cdd.drdmyanmar.org/en> Accessed 15.05.19.

National Community Driven Development Project. *NCDDP management information system.* (2018b). <https://www.ncddmis.com/gis_y6/index.php> Accessed 14.05.19.

Nemoto, K. (2014). *Monogatari Biruma no rekishi (in Japanese).* Tokyo: Chuokoron-shinsha, Inc.

Numata, M., Sugiyama, M., & Mogi, G. (2018). *Barrier analysis for deployment of minigrids in Myanmar (in Japanese). 37th annual meeting of Japan Society of* energy and resources (pp. 222–225). Osaka: Japan Society of Energy and Resources.

Numata, M., Sugiyama, M., & Mogi, G. (2021). Distributed power sources to improve the decent living standard (DLS) in the ethnic minority areas of Myanmar. *Sustainability, 13*(6), 3567. Available from https://doi.org/10.3390/su13063567.

Numata, M., Sugiyama, M., Mogi, G., Wunna S., & Anbumozhi, V. (2018). Technoeconomic assessment of microgrids in Myanmar (ERIA Discussion Paper Series No. ERIA-DP-2018-05). Jakarta. <http://www.eria.org/publications/technoeconomic-assesment-of-microgrids-in-myanmar/>.

Nyein, N. *Analysis: Why did the KNU temporarily leave peace talks?* (2018a). <https://www.irrawaddy.com/factiva/analysis-knu-temporarily-leave-peace-talks.html> Accessed 20.11.18.

Nyein, N. *Tatmadaw announces four-month ceasefire in north, northeast.* (2018b). <https://www.irrawaddy.com/news/tatmadaw-announces-four-month-ceasefire-north-northeast.html> Accessed 25.12.18.

Office of the Civil Service Commission. *Myanmar today.* (n.d.). <https://www.ocsc.go.th/sites/default/files/attachment/page/myanmar.pdf> Accessed 26.12.18.

Pact. *Myanmar – pact.* (2018). <https://www.pactworld.org/country/myanmar/project> Accessed 28.05.19.

Parks, T., Colletta, N., & Oppenheim, B. (2013). The contested corners of Asia: Subnational conflict and international development assistance. The Asia Foundation. San Francisco. <https://asiafoundation.org/resources/pdfs/ContestedCornersOfAsia.pdf>.

Du Pont, P. (2019). Decentralizing power:The role of state and region governments in Myanmar's energy sector. Yangon. <https://asiafoundation.org/wp-content/uploads/2019/04/Myanmar-Decentralizing-Power_report_11-April-2019.pdf>.

Pwint, N.L. H. *Arakan army chief promises Myanmar military, govt eye for an eye.* (2019). <https://www.irrawaddy.com/in-person/arakan-army-chief-promises-myanmar-military-govt-eye-eye.html> Accessed 21.01.19.

Raleigh, C. *Myanmar: Conflict update.* (2018). <https://www.acleddata.com/2018/04/05/myanmar-conflict-update/> Accessed 04.04.19.

Republic of the Union of Myanmar. Constitution of the Republic of the Union of Myanmar (2008). Myanmar. <http://www.burmalibrary.org/docs5/Myanmar_Constitution-2008-en.pdf>.

Saferworld, & Karen Peace Support Network. (2019). Security, justice and governance in South East Myanmar: A knowledge, attitudes and practices survey in Karen ceasefire areas. London and Yangon. <https://www.saferworld.org.uk/resources/publications/1194-security-justice-and-governance-in-south-east-myanmar-a-knowledge-attitudes-and-practices-survey-in-karen-ceasefire-areas>.

Salem-Gervais, N., & Raynaud, M. *Ethnic language teaching's decentralisation dividend.* (2019). <https://frontiermyanmar.net/en/ethnic-language-teachings-decentralisation-dividend> Accessed 29.05.19.

Shida, J. (2017). Japan's Realism and Liberalism (No. 6). Sasakawa USA Forum. Washignton DC. <https://spfusa.org/wp-content/uploads/2017/03/Shida-Japan-realism-liberalism.pdf>.

Simpson, A., & Smits, M. (2018). Transitions to energy and climate security in Southeast Asia? Civil society encounters with illiberalism in Thailand and Myanmar. *Society and Natural Resources, 31*(5), 580–598. Available from https://doi.org/10.1080/08941920.2017.1413720.

Smith, M. (1994). Ethnic groups in Burma. (A.-M. Sharman, Ed.), Anti-Slavery international. London: Anti-Slavery International. <http://www.ibiblio.org/obl/docs3/Ethnic_Groups_in_Burma-ocr.pdf>.

South, A., Schroeder, T., Jolliffe, K., Non, M.K. C., Shine, S., Kempel, S., Mu, N.W. S. *Between Ceasefires and federalism: Exploring interim arrangements in the Myanmar peace process.* (2018). <https://covenant-consult.com/wp-content/uploads/MIARP-Report.pdf> Accessed 07.02.19.

Sovacool, B. K. (2012). The political economy of energy poverty: A review of key challenges. *Energy for Sustainable Development, 16*(3), 272–282. Available from https://doi.org/10.1016/j.esd.2012.05.006.

Sovacool, B. K., Heffron, R. J., McCauley, D., & Goldthau, A. (2016). Energy decisions reframed as justice and ethical concerns. *Nature Energy, 1*(May), 1–6. Available from https://doi.org/10.1038/nenergy.2016.24.

Takahashi, A. (2018). *Sosei no Myanmar (in Japanese).* Tokyo: Akashi shoten.

Tang, K., & Zhao, Y. (2017). Health as a bridge to peace and trust in Myanmar: The 21st century Panglong conference. *Globalization and Health, 13*(40), 1–4. Available from https://doi.org/10.1186/s12992-017-0271-3.

The Irrawaddy *Analysis: Arakan Army - A Powerful New Threat to the Tatmadaw.* (2019). <https://www.irrawaddy.com/news/burma/analysis-arakan-army-powerful-new-threat-tatmadaw.html> Accessed 10.05.19.

The Nationwide Ceasefire Agreement Between the Government of the Republic of the Union of Myanmar and the Ethnic Armed Organisations. (2015). <https://peacemaker.un.org/sites/peacemaker.un.org/files/MM_151510_NCAAgreement.pdf> Accessed 05.04.19.

The Republic of the Union of Myanmar. *Myanmar's intended nationally determined contribution-INDC.* (2015). https://www4.unfccc.int/sites/ndcstaging/PublishedDocuments/Myanmar First/Myanmar%27s INDC.pdf.

Thit, M. *WWF-Myanmar to support renewable energy in Kayin State.* (2018). <http://www.globalnewlightofmyanmar.com/wwf-myanmar-to-support-renewable-energy-in-kayin-state/> Accessed 28.05.18.

Tun, Y.M. *Ethnic data from 2014 census to be released.* (2017). <https://www.mmtimes.com/national-news/24393-ethnic-data-from-2014-census-to-be-released.html> Accessed 19.12.18.

Tønnesson, S., Aung, N.L., & Nilsen, M. (2019). Will Myanmar's Northern Alliance join the peace process ? (PRIO POLICY BRIEF). Oslo. <https://www.prio.org/utility/DownloadFile.ashx?id = 1757&type = publicationfile>.

Tønnesson, S., & Nilsen, M. (2018). Still a chance for peace in Myanmar? (No. 2). PRIO Policy Brief. Oslo. <https://www.prio.org/utility/DownloadFile.ashx?id = 1562&type = publicationfile>.

Union of the Republic of Myanmar. National Education Law, Pub. L. No. 1376 (2014). Union of the Republic of Myanmar. <http://www.burmalibrary.org/docs20/2014-09-30-National_Education_Law-41-en.pdf>.

United Nations. *Polio Immunisation crosses conflict borders in Myanmar.* (2016). <http://mm.one.un.org/content/unct/myanmar/en/home/news/polio-immunisation-crosses-conflict-borders-in-myanmar-.html> Accessed 29.05.19.

United States Institute of Peace. (2018). China's role in Myanmar's Internal Conflicts. Washington, DC. <https://www.usip.org/publications/2018/09/chinas-role-myanmars-internal-conflicts>.

University Of Yangon. *History: Yangon university.* (n.d.). <https://www.uy.edu.mm/yangon-university/> Accessed 28.05.19.

Visser, L.J. (2016). Building relationships across divides: Peace and conflict analysis of Kachin State. (T. Ma, Ed.). Centre for peace and conflict studies. <https://reliefweb.int/sites/reliefweb.int/files/resources/Conflict-Analysis-Kachin-State-29.9.16.pdf>.

Wallechinsky, D. *The world's 10 worst dictators.* (2009). <https://parade.com/110573/davidwallechinsky/the-worlds-10-worst-dictators/> Accessed 15.06.19.

Weng, L. *More villagers flee fighting between rival armed groups in Northern Shan.* (2018). <https://www.irrawaddy.com/news/burma/villagers-flee-fighting-rival-armed-groups-northern-shan.html> Accessed 28.12.18.

Weng, L. *Nationwide ceasefire agreement no Panacea for what Ails Myanmar.* (2019a). <https://www.irrawaddy.com/news/burma/bangladesh-govt-to-hold-jobs-fair-for-coxs-bazar-youth-following-protests.html> Accessed 25.04.19.

Weng, L. *RCSS invites Rival Shan group to join ceasefire, excludes TNLA.* (2019b). <https://www.irrawaddy.com/news/burma/rcss-invites-rival-shan-group-to-join-ceasefire-excludes-tnla.html> Accessed 10.05.19.

World Bank. *National Community Driven Development Project Frequently Asked Questions.* (2016). <http://www.worldbank.org/en/news/feature/2014/03/31/qa-myanmar-national-community-driven-development-project#6> Accessed 21.11.18.

World Bank. (2019). Disclosable Version of the ISR - National Electrification Project - P152936 - Sequence No: 07 (English). Washignton, DC. <http://documents.worldbank.org/curated/en/153761548028343612/pdf/Disclosable-Version-of-the-ISR-National-Electrification-Project-P152936-Sequence-No-07.pdf>.

WWF. *The Barefoot project.* (2018). <http://www.wwf.org.mm/en/the_barefoot_project/> Accessed 14.12.18.

XE.com Inc. *JPY to USD.* (2019). <https://www.xe.com/currencyconverter/convert/?Amount = 495%2C000%2C000&From = JPY&To = USD> Accessed 23.05.19.

Yangon Technological University. *Historical background.* (n.d.). <https://ytu.edu.mm/page/7> Accessed 28.05.19.

Yun, L. (2017). Building peace in Myanmar: Birth of the FPNCC. *Asia Dialogue*, 2016–2018. Available from https://doi.org/10.5271/sjweh.3544.

6

Energy development for the stateless: Rohingya case study

Samira Siddique
Energy and Resources Group, Renewable and Appropriate Energy Laboratory, University of California, Berkeley, CA, United States

Chapter outline

6.1 Introduction

In August 2017 the news media were gripped by images of the world's most recent refugee crisis. Photos of Rohingya people—often dubbed the world's most persecuted minority—showed families fleeing across the Naf River from Myanmar into Bangladesh with the smoke of their burning villages at their backs. In a span of three months, over 800,000 Rohingya fled into Bangladesh, settling in an area of approximately 23 square kilometers outside of Cox's Bazar, Bangladesh. As an ethnic Muslim minority group, the Rohingya were escaping decades-long persecution in Myanmar. The persecution reached an apex when a small group of Rohingya militants attacked a Myanmar police post, resulting in violent retaliatory attacks by extremist Buddhist groups and the military (Beech, 2017).

As the Rohingya influx was covered in the news, there was little information provided about the history of this group of people. For example, Rohingya people had been living in this area of Bangladesh for decades, some having fled Myanmar

Energy Policy for Peace. DOI: https://doi.org/10.1016/B978-0-12-817350-3.00001-8

during earlier retaliatory periods. Using the term "crisis," as in the "Rohingya Crisis," suggest a beginning and end to a challenging situation. In reality, protracted displacement situations, as is the case for Rohingya who have resided in Bangladesh since the 1970s, last an average of 26 years (UNHCR, 2018). There are people who are born and grow up in refugee camps and then raise their own families there. All the while, they are stateless and are not given basic political and economic rights.

Like most other refugee camp situations, the Rohingya camp will likely last for at least another decade. Therein lies the problem: Refugee situations are not conceptualized as long-term situations. This affects the extent of aid that is given, the type of facilities that are built, and of course the economic and political rights and social support that the Rohingya have.

Over the course of a few months in mid-2018, I spoke to United Nations (UN) and International Organization for Migration (IOM) aid workers, project officers, nongovernment organization (NGO) workers, and other researchers working on Rohingya issues in Bangladesh for preliminary fieldwork. In this time span, there was a palpable shift from short-term emergency relief to longer-term planning within the Rohingya camps. The UN and IOM officers with whom I spoke alluded to infrastructure development projects, and many NGO workers were also beginning to adopt a development mindset. A good example of this shift is that at the start of the summer, there were no funded projects around energy access. By the end of the summer, the World Bank and the Development Bank (ADB), two of the largest development fund agencies in the world, had set aside funds specifically for solar energy in the Rohingya camps.

The delicate politics of the situation and perhaps the history of the aid industry have made it rare to find anything about "development" in the refugee and migrant literature. There is an inherent tension between refugee camps and migrant settlements and the idea of development. Encampments are conceptualized as temporary dwellings, whereas the objective of development (political, economic, social, infrastructure, etc.) is to create sustainable communities for the long term. Being in the field in the Rohingya camps and my conversations with people working there have alerted me to the need to research this emerging phenomenon.

Refugee crises and mass migrations will continue to happen at varying scales, whether through ethnic cleansing, environmental disaster, economic crisis, or something else. The UN, development agencies, NGOs, and some governments are only just beginning to rethink how we prioritize refugees and migrants and integrate them into existing development frameworks. But we will likely have to think up a whole new development paradigm for those that have been systematically "othered" and persecuted.

This study examines the shift from short-term emergency relief efforts to development planning in displacement settings via energy access implementation through a literature review that spans a variety of social science disciplines. First, I discuss more in depth the interdisciplinary nature and methods that I used for my literature review, which was necessary for me to establish the development-displacement research space. Second, using the theoretical idea of "bare life" from refugee and forced migration studies, I suggest that the shift from emergency relief to development necessitates a critical lens in viewing the hierarchical nature of service provision in refugee camps. Third, I establish energy access and implementation projects as a prime example of a shift toward development in displacement settings, and I explore the research and aid investments around this issue. Finally, I ground this literature review with my own observations during preliminary fieldwork in the Rohingya refugee camps, where I observed firsthand as this shift from emergency relief to development took place.

I do not intend to make normative claims about whether it is "right" to approach planning in the refugee camp setting from a development perspective. In fact, not enough monitoring and evaluation studies have been done in displacement settings to make this claim. So questions such as "What constitutes good development in a displacement setting?" are currently unanswerable. I do, however, believe that it is critical to think about how best to approach long-term planning for stateless people. As the number of displaced people worldwide continues to grow, it is critical to think deeply about how to implement social development and service provision in a way that is sustainable for these highly vulnerable populations. With over 68.5 million displaced people worldwide (UNHCR, 2018), this population is likely to become increasingly relevant in development studies. It will be important for researchers, policymakers, practitioners, and aid workers alike to perform the intellectual exercise of conceptualizing what development for stateless people should look like in practice.

6.2 Energy access and development literature

To understand the development implementation aspect of this research further, I look at the energy access literature. I choose to focus on energy access because I view it as an entry point into the development field in displacement settings as a whole. Unlike other sectors, such as water, sanitation, and hygiene (WASH), energy is not seen as an immediate need in emergency relief settings. Thus newer innovations and greater financing for longer-term energy projects in displacement settings, as opposed to those in WASH, can arguably be seen as a move toward more development.

There has been a push among the monitoring and evaluation research community to study the long-term implications of development projects since the establishment of the United Nations Sustainable Development Goals (SDG), a set of 17 interconnected sustainable development goals for "people, planet, and prosperity" by 2030, which is intended to guide countries in their national development action plans. SDG 7 is to ensure access to affordable, reliable, sustainable, and modern energy for all. In a recent paper in *Nature* by Nerini et al. (2018), this goal was analyzed to have the most synergies with all the other goals and the fewest tradeoffs in terms of economic development for most countries, resulting in a specific push for energy modelers to study alternative energy scenarios further in developing countries. However, there is still a dearth of this energy modeling research in displacement settings.

There is some literature in this space from energy geographers, but these studies are not concerned with displacement settings. There is also a growing literature from more quantitative energy modelers, but they are not at all community-based in their approach. Studies that analyze the implementation and feasibility of energy service provision in displacement settings are limited to older refugee camps, such as those in Palestine or Syria. Not enough monitoring and evaluation studies have been done to test the usefulness of decentralized energy experiments, such as solar minigrids, and more research is needed in this area.

6.3 "Bare life" and the relief-to-development gap

It is impossible to study the shift from emergency relief to development without bringing up an oft-studied trope in refugee and migration studies, that of "bare life." This is an idea that was first proposed by the philosopher Giorgio Agamben, highlighting the prioritization of a refugee's most primitive needs by host states and the aid industry above any individual political or social voice (Agamben, 1998). It HAS BEEN suggested that "bare life refers to human life in which the mere biological fact of life is prioritized over the way a life is lived—when the prospects of life, with all of its potential, possibilities, and forms, are reduced to sheer biological life and are abandoned to the unconditional power of the sovereign. The [refugee] camp ... is where bare life is produced" (Katz, 2017).

The way in which humanitarian aid programs are implemented perpetuates this notion of bare life—that only the most immediate or "biological" needs of the refugee are tended to because they exist outside of a formal state regime. Turner breaks this idea down further:

> To be worthy of humanitarian assistance, the receiver must be purely human—that is someone without a past, without political will, without agency ... the refugee can no longer voice his political rights but rather appeal to a common humanity by showing his wounds (Turner, 2016).

The infrastructure of refugee camps often reinforces the idea of bare life, sometimes being used for the surveillance of refugees when the camp is spatially constructed to be hyperorganized and gridlike and creating barriers between the refugees and host community and aid workers. As Jaji (2012) notes,

> The use of geography to facilitate control of refugee movement and interaction with those outside the camp is mirrored by the physical layout of the camp. Unlike the surrounding villages, which are mostly haphazard, the 'architecture' of the two camps is determined by the quest for control, discipline and order. ... The layout, petty as it may seem, is a powerful surveillance and disciplinary mechanism which 'renders visible those who are inside [the camp]' (Jaji, 2012).

Often, the camp hierarchy in itself can enable the relief-to-development gap. The UN Refugee Agency (UNHCR) cannot work against the host government and therefore often restricts the movement of refugees outside their encampments. Humanitarian organizations also often recreate the same emergency relief processes that they used in previous encampment situations. To this extent, the processes may not be appropriately contextualized to the needs of the refugees, and they may be difficult to build upon in transitioning to longer-term planning. Hierarchy is

> reinforced by the control and maintenance of social distance between camp administration and aid agencies on the one hand and refugees on the other hand. This social distance has a physical and conspicuous manifestation in the form of fencing and fortressing of the UNHCR compounds. These physical

barriers denote the communication barrier between humanitarian organizations' staff and refugees; the staff are simultaneously available and conspicuous but inaccessible, physically near but socially distant (Jaji, 2012).

The distance between refugees and aid workers makes it difficult to overcome the top-down nature of projects that are implemented inside camps and tends to close off the refugee population even further from the outside community.

As there is an increasing shift toward longer-term development in refugee camps and perhaps a shift away from this idea of bare life toward more sustainable solutions, it is critical to address the fundamentally hierarchical nature of these institutions and spaces. One of the most explicit ways in which there has been a possible shift away from bare life in displacement settings is the increased investment by aid and development agencies in service provisions and longer-term projects that have traditionally not been made in refugee camps. A prime example of this is the increase in investments in alternative energy sources in displacement settings. Energy has historically not been considered an emergency relief priority in refugee camps but has become a larger priority for aid agencies as protracted displacement situations have become more prevalent globally.

6.4 Energy provision and research in refugee camps

With the adoption of SDG 7 in September 2015, the international community established energy as a fundamental pillar of development for the first time (Lehne, Blyth, Lahn, Bazilian, & Grafham, 2016). Despite setting this goal in the development sphere, there has been relatively little focus on energy issues in humanitarian settings (Energy Access Practitioners Network, 2018). Energy needs are fundamental to the basic needs of displaced people, from cooking, heating, and water pumping to medical care and communications. But historically, there has not been a systematic approach to energy supply in conflict settings because they are thought to be shorter term. Most of the energy is supplied ad hoc by individual NGOs or international aid agencies, usually through the use of diesel generators. In part, this is due to a lack of available funding for energy services. In contrast, other basic needs, such as food, water, sanitation, shelter, and health, have been the focus of aid agencies, practitioners, and researchers. The inadequate energy supply measures that are introduced as temporary stopgap measures in emergency circumstances can, over time, lead to expensive, unhealthy, and inefficient processes.

UNHCR, development aid agencies, energy access practitioners, and researchers are beginning to take notice of the energy gap that exists in displacement settings. They are each addressing this gap within their means, but there exists a divide between policy, research, and project implementation.

6.4.1 UNHCR and aid agencies explore alternative energy solutions

In recent years, the role of UNHCR and other humanitarian agencies has changed significantly. Originally, these institutions were focusing on the provision of short-term, temporary emergency relief and operated under the premise that forcibly displaced people may eventually return home. However, statistics show that forcibly displaced people can remain in refugee camps for an average of 26 years, which means that managing the camps is more often than not a longer-term obligation that requires a dedicated strategy.

The topic of energy demand and supply in refugee camps has gained increasing attention from the aid industry in recent years. UNHCR, energy practitioners, and other actors have launched international working group initiatives, such as the Global Strategy for Safe Access to Fuel and Energy, the Moving Energy Initiative, and the Energy Access Practitioners Network, to explore lower-cost energy options in displacement settings. At the same time, a variety of different pilot projects funded by private foundations and development fund agencies have started to bring clean energy technologies, such as solar photovoltaic and solar minigrids, to the field and to position the humanitarian sector as a launch pad for sustainable development. Examples include the distribution of solar lanterns, which allow camp inhabitants to light their shelters or charge mobile phones, and the installation of solar street lights, which improve camp security and allow for communal activities after nightfall (IKEA Foundation, 2016). In addition to small-scale, decentralized solar solutions, the recent inauguration of two megawatt-scale photovoltaic power plants near the refugee settlements of Azraq and Mafraq Za'atari in Jordan demonstrate that solar photovoltaic systems can be used to power vital infrastructure of entire camps (Ossenbrink, Pizzorni, & van der Plas, 2018).

Historically, development fund agencies, such as the World Bank and the ADB, have not funded humanitarian relief efforts. This has changed in recent years, as such agencies have developed projects to assist both refugee and host communities in developing countries where they are already based (The World Bank, 2017). In an unprecedented effort, the ADB pledged US$2 million to provide solar minigrid systems to the Rohingya refugee camps in Bangladesh and surrounding host communities by mid-2019—the first energy investment of this scale from a development fund agency.

Top-down project implementation plans, including the one proposed by the ADB, are common in displacement settings and typically lack community-based needs assessments. Moreover, there is a severe lack of knowledge regarding how the political status of refugees will hinder their ability to benefit from energy infrastructure. Although there is great potential in alternative energy technologies, as has been proven by the many studies on alternative energy that have been conducted in other settings, more longer-term monitoring and evaluation research are needed.

6.4.2 Energy modeling research for alternative energy solutions

As SDG 7 stipulates, "energy access for all" has become a driving force for energy optimization modelers and researchers who work in developing countries. Many of the energy optimization models project cost savings and the feasibility of setting up decentralized renewable energy in emerging economies. The thought is that transitioning, or starting out with, more sustainable energy sources before an economy is too dependent on dirty fuels could result in more equitable access to energy as well as a significant reduction in global carbon emissions (Lovins, 2016). There is a small but growing literature on energy modeling, often with renewable energy alternatives, in displacement settings. The modeling methods add an important quantitative dimension to research and project implementation in displacement settings. It shows that there are other more dependable, cost-saving, and sustainable energy alternatives to diesel, such as solar minigrids, and they give decision makers concrete numbers with which to implement suitable energy projects and contribute much needed data on an understudied area.

Despite growing work in energy modeling in displacement settings, there exists a paucity of high-quality data on energy access, demand, and need. There are few centralized systems for gathering data on this issue and few existing studies that gather data beyond single cases. Most of the research into energy in displacement settings has focused on the impact of improved cookstoves on fuel wood use (Lehne et al., 2016). While limited, the existing literature on non-cookstove interventions in displacement settings covers a broad range of alternative and renewable energy solutions. Nerini et al. (2015), for example, studied the potential for a portable module to sustainably meet the energy and water needs of displaced populations in protracted emergency settings. Naso, Micangeli, and Michelangeli (2013) studied the effect of solar thermal collectors during the postemergency phase after the Abruzzi earthquake.

In one of the first alternative energy modeling studies in displacement settings, Lehne et al. modeled energy demand with an end-use accounting approach, arriving at global demand estimates by scaling up from data on energy use patterns at the household level using global datasets from UNHCR (Lehne et al., 2016). The study models three different alternative energy scenarios. In the first scenario, all displaced households retain their baseline patterns of energy use for cooking but do so more efficiently as well as adopting basic solar lanterns and household diesel generators to meet the most basic lighting needs. In the second scenario, all households that were previously dependent on solid biomass for cooking shift two-thirds of their fuel consumption to biomass briquettes. In terms of lighting, the households that were previously dependent on torches and kerosene adopt a 50/50 split between solar minigrids and solar lanterns. In the third scenario, all displaced households adopt liquid petroleum gas as their main cooking fuel. In terms of lighting, the majority of households again adopt a 50/50 split between solar and minigrid solutions but with a higher level of access. The first scenario would be easiest to achieve because the upfront costs of the technology are relatively low and the annual fuel savings for displaced people would be considerable. The second scenario would generate similar gains in lower emissions but would cost more in both capital and fuel costs. The third scenario is the most expensive but would yield huge potential benefits in terms of safety, health, protection, and market generation.

Studies such as those of Lehne et al. (2016) are providing important data for baseline energy needs and potential alternative energy scenarios in displacement settings. They show that energy should be incorporated as a key consideration at each stage of a humanitarian response and that planning for energy provision should be a feature of the immediate emergency response and then be revisited and updated as the situation develops. They also demonstrate the need for a systematic approach to collecting and reporting data on fuel use, energy practices, and costs. In doing so, aid agencies and policymakers have the tools to explore new and longer-term funding and delivery models for energy in displacement settings and humanitarian relief more generally. The current short-term funding model does not lend itself to investments in alternative energy solutions in which the savings, as indicated by alternative energy modeling analysis, are likely to accrue after a number of years.

However, these modeling studies are still limited in capturing the scope of potential social benefits and possible limitations to the displaced population themselves. Lehne et al. (2016) and similar studies conducted interviews solely with UNHCR representatives, which is not emblematic of the energy needs of the community that is receiving the new technologies. While baseline data are important for modeling projections, it is also necessary to collect iterative monitoring and evaluation data to understand whether these technologies turn out to be as beneficial as they are projected to be.

Much work has been done by social scientists, particularly ethnographers, in displacement settings. There is also a reemerging literature on the social, economic, and spatial aspects of energy infrastructure transitions, particularly by geographers (Bridge, Bouzarovski, Bradshaw, & Eyre, 2013). However, there has not been much social science research conducted on the implications of new energy technologies in displacement settings. Refugee camps are massive enterprises with complex social, political, and spatial dynamics (Nadai & van der Horst, 2010). For the world to meet the UN Sustainable Development Goals, it is critical to understand the social impact of energy access in tandem with the modeling work and top-down project implementation in these settings.

6.5 Energy access and development in the Rohingya refugee camps

There are currently upwards of 1 million Rohingya refugees living in Cox's Bazar, Bangladesh. It is the largest and most recent refugee camp in the world. Although there have been ongoing discussions between the Bangladeshi government and UNHCR to relocate the Rohingya to a remote island off the coast of Bangladesh or repatriate them to Myanmar, it is almost certain that they will remain where they are for an indefinite period of time. Like most other refugee camp situations, this one will likely last for at least another decade. Many NGOs and aid agencies that are working on Rohingya issues realize that this is not temporary and are beginning to take a longer-term view of the camps. The shift from emergency relief to development has begun, underscoring the fact that the refugee crisis has huge long-term implications for how development operates beyond state citizenship.

A strong indication of the shift toward development in the Rohingya camps are the recent investments by the World Bank and the ADB, of $480 million and $100 million, respectively (The World Bank, 2018). Traditionally, these two institutions have invested in long-term development projects and have supported governments in capacity building. In the past few years, they have created a relief fund for emergency situations such as the Rohingya crisis.

One of the investments from the World Bank and the ADB is in renewable energy in the Rohingya camps. The investment in energy access shows a gradual shift toward longer-term or at least medium-term planning in the camps. Compared to international aid funding in every other sector—water, sanitation, health, shelters, and so on—energy had no allocated funding at the beginning of the Rohingya influx. This was largely because energy is not seen as essential to emergency relief, which is arguably an outdated view from the aid industry, as energy access is linked to more positive health effects and gender safety and equality (Kammen & Dove, 1997). With the World Bank and ADB's investment plan, a portion was allocated to set up solar minigrids in 2019 as well as constructing more solar lamps and distributing solar lanterns. The move toward renewable energy shows increasing interest in long-term development because it is inherently sustainable and simple to use. A solar minigrid offers a cleaner and more consistent alternative to diesel generators and can potentially be used to anchor local minigrids if the refugee camps are present in the longer term (Mozersky & Kammen, 2018). Additionally, solar lighting in the camps can provide an element of safety and security for a population that consists predominantly of women and children.

Of all the Rohingya camps, it is striking that the only one that is connected to the national electricity grid, and thus is situated for the longer term, is a camp in Teknaf, where some Rohingya have lived for many years and have essentially been assimilated into the surrounding community. Perhaps the thought is that there is "value added" if the Rohingya contribute economically, so it makes sense to invest in electricity lines. However, this situation is exceedingly rare, as the vast majority of Rohingya cannot move freely outside the camps and therefore are unable to be economically independent.

While the notion of development is important for improving livelihoods, the development itself must be done differently for the stateless. Traditional forms of economic development do not work for stateless people who have no means to gain employment. Although some cash-for-work programs and recreation facilities have been set up by aid agencies, the vast majority of Rohingya have nothing to do during the day; their routines are often set around food and aid distribution schedules. They are recovering from unimaginable trauma. The camps will only continue to grow: Rohingya are still crossing the border, though at much lower rates, and there are projected to be 50,000 babies born this year. No amount of aid distribution or traditional notions of development will fix these facts of life for the Rohingya.

Part of the difficulty in streamlining development efforts is the institutional power structure of the camps. After the exodus began in August 2017, the Bangladeshi government did not officially labeled the Rohingya as "refugees." Without this label, UNHCR could not head the emergency relief operations in the camps, as they normally would when refugees are involved. Therefore the IOM took over camp operations. Within a few months, UNHCR was allowed to work in the camps, and it started coleading operations with IOM. The two humanitarian stakeholders now oversee relief operations in about ten sectors and work alongside the government's response to the crisis, which includes different government agencies and the Bangladeshi army.

Perhaps in response to the historical conflicts between host and refugee communities, aid agencies based in the Rohingya camps are implementing their longer-term projects for both the host and Rohingya populations. A long-term environmental monitoring project, carried out by the International Centre for Climate Change and Development contracting under the International IOM, looks at the changes in water quality and air pollution from cooking fuel in both the Rohingya and host communities over time (ISCG, 2018). Social services for women and children are also available to both communities, such as the recent investment of $100 million from the Lego Foundation to Sesame Workshop to bring "learning through playtime" spaces to Syrian and Rohingya refugees and host communities (BRAC USA, 2018).

The future of the Rohingya in Bangladesh remains unclear. There have been more recent efforts by the Bangladeshi government to repatriate the Rohingya to Myanmar, but the conditions were deemed unfit by the UNHCR, and the Rohingya themselves were unwilling to repatriate because of the poor conditions that awaited them in Rakhine state (Uddin, 2018). Additionally, concrete, gridlike shelters are being built for approximately 100,000 Rohingya to relocate to Bashan Char Island off the coast of Bangladesh, but it remains uncertain how and when the Rohingya would be relocated there (The Guardian, 2018). While these plans remain uncertain, the medium- to longer-term development projects in the Rohingya camps were slated to begin in 2019.

6.6 Conclusion: future implications for the energy development in displacement settings

My study suggests that multiscale and multidisciplinary approach with a focus on community-based methods is the best way to approach the analysis of development projects in displacement settings. Because there is an increased focus on development within the humanitarian regime, researchers, policymakers, advocates, and practitioners must think critically about how to integrate stateless people or perhaps shift to a new development paradigm entirely.

A major challenge is that there is no direct guiding principle globally for how to integrate stateless people, let alone how to develop communities with them in mind. For example, none of the UN Sustainable Development Goals explicitly address development for stateless persons. International NGOs and the UN could adopt a more explicitly rights-based approach to development, especially because more refugee crises and mass migrations are projected to occur in the future. This approach would combine different existing concepts of international development, such as capacity building, human rights, participation, and sustainability (Nussbaum, 1998). The goal would be to empower the group that cannot exercise full rights and to strengthen the capacity of institutions and governments that are obligated to fill these rights.

However, the main criticism of the rights-based approach is that it merely incorporates the language of human rights with development but does not change the programs that are being implemented (Nelson, 2007). For change to take place, governments must be willing to accept refugees and migrants and hold other countries accountable for the processes that lead to refugees in the first place. Many governments that receive refugees, whether willingly or not, are not capable of developing long-term communities for the refugees in their own country. If development becomes a new appendage to refugee and migration studies or if the field of development studies itself casts an eye toward displaced populations, this critical thought is necessary from both a policy and research point of view.

More broadly, there is a gap in evaluation of basic service provisions in displacement settings, prompting a reliance on evidence generated by a small number of postintervention, observational studies. The timeline of the ADB's solar minigrids implementation in the Rohingya refugee camps provides a unique opportunity to conduct a well-controlled before-and-after evaluation. Implementation projects such as ADB's necessitate a community-based angle that is often missing from energy implementation projects, especially those in contexts with hierarchical power structures such as refugee camps.

As the number of displaced people worldwide continues to grow and as aid shifts toward longer-term implementation efforts, there is a need to understand how these particular political, geographic, and social contexts should influence the aim and practice of development whether through research, policy, advocacy, or all three.

References

Agamben, G. (1998). *Homo sacer.* Stanford University Press.

Beech, H. (2017). Desperate Rohingya flee Myanmar on trail of suffering: 'It is All Gone'. The New York Times, September 2.

BRAC USA. (2018). *The LEGO foundation awards $100 million to Sesame workshop.* USA: BRAC.

Bridge, G., Bouzarovski, S., Bradshaw, M., & Eyre, N. (2013). Geographies of energy transition: Space, place and the low-carbon economy. *Energy Policy, 53*(1), 332.

Energy Access Practitioners Network. (2018). Social impact of mini-grids: Monitoring, evaluation and learning tools and guidelines for practitioners and researchers. *Energy Access Practitioners Network.*

IKEA Foundation. (2016). *Annual review 2016: Helping families most at risk from climate change.* IKEA Foundation.

ISCG. (2018). *IOM Bangladesh: Rohingya* humanitarian crisis res*ponse*—external up*date (27 July—2 August 2018).* Intersector Coordination Group.

Jaji, R. (2012). Social technology and refugee encampment in Kenya. *Journal of Refugee Studies, 25*(2), 224.

Kammen, D., & Dove, M. (1997). The virtues of mundane science. *Environment, 39*, 6.

Katz, I. (2017). Between bare life and everyday life: Spatializing Europe's migrant camps. *Architecture_MPS, 12*(2), 2.
Lehne, J., Blyth, W., Lahn, G., Bazilian, M., & Grafham, O. (2016). Energy services for refugees and displaced people. *Energy Strategy Reviews, 12–14*, 131.
Lovins, A. (2016). Soft energy paths: Lessons of the first 40 years. *Solutions The Journal, 9*, 1.
Mozersky, D., & Kammen, D. (2018). *South Sudan's renewable energy potential: A building block for peace.* United States Institute of Peace.
Nadai, A., & van der Horst, D. (2010). Introduction: Landscapes of energies. *Landscape Research, 35*(2), 144.
Naso, V., Micangeli, A., & Michelangeli, E. (2013). Sustainability after the thermal energy supply in emergency situations: The case study of Abruzzi earthquake. *Sustainability, 5*(8).
Nelson, P. (2007). Human rights, the millennium development goals, and the future of development cooperation. *World Development, 35*(12).
Nerini, F., et al. (2018). Mapping synergies and trade-offs between energy and the sustainable development goal. *Nature Energy, 3*(1).
Nerini, F., Valentini, F., Modi, A., Upadhyay, G., Abeysekara, M., Salehin, S., & Appleyard, E. (2015). The energy and water emergency module; A containerized solution for meeting the energy and water needs in protracted displacement situations. *Energy Conversion and Management, 93*, 206.
Nussbaum, M. (1998). Capabilities and human rights. *Fordham Law Review, 66*(2).
Ossenbrink, J., Pizzorni, P., & van der Plas, T. (2018). Solar PV systems for refugee camps: A quantitative and qualitative assessment of drivers and barrier.
The Guardian. (2018). Concrete camps being built for Rohingya refugees in Bangladesh. *The Guardian.*
The World Bank. (2017). *Additional financing available to support refugees and host communities.* The World Bank.
The World Bank. (2018). *The world bank announces support for Bangladesh to help Rohingya.* The World Bank.
Turner, S. (2016). What is a refugee camp? Explorations of the limits and effects of the camp. *Journal of Refugee Studies, 29*(2), 143.
Uddin, N. (2018). Ongoing Rohingya repatriation efforts are doomed to failure, November 22 *Al Jazeera.*
UNHCR. (2018). *Figures at a glance.* United Nations Refugee Agency.

7

Toward inclusive and sustainable rural energy transition: defining parameters of successful community participation in India

Venkatachalam Anbumozhi
Economic Research Institute for ASEAN and East Asia, Jakarta, Indonesia

Chapter outline

7.1 Introduction

Biomass, such as fuel wood and agricultural residues, is still the dominant source of fuel and contributes major parts of total cooking energy in many developing countries in Asia. In rural India, women and girl children are the key players in

Energy Policy for Peace. DOI: https://doi.org/10.1016/B978-0-12-817350-3.00002-X

producing, collecting, and using these fuels. However, inefficient use of such fuels is a factor in air pollution and health hazards besides causing economic loss to the nation. Another major cause of concern has been the indiscriminate use of fuel wood, leading to deforestation and desertification. In this context, technological solutions, institutional arrangements, awareness creation, and training schemes for ensuring adequate affordable clean energy systems and services assume great significance in rural development and energy policy programs. Many rural energy programs in India have been designed to increase energy supplies to rural households in a sustainable manner through the introduction of renewable energy technologies. Their focus has been on improving energy use efficiency in cooking and lighting devices. To bring sustainability to such energy transition programs, decentralized participatory methods are necessary for planning, implementation, and project monitoring (John & Chathukulam, 2001). The emphasis on a participatory approach stems because energy intervention cannot be seen in isolation from the wider rural development priorities of the people (Norman, 1978). Rural energy issues or patterns of consumptions are localized; therefore the strategies to tackle these issues need to be planned at the local level (Reddy & KrishnaPrasad, 1977). Moreover, the capacity of the people as well as capability of the community is valuable in understanding the introduction of new development systems and identifying workable solutions (Haq, 1976). The participation of the village community needs to be sought at various levels and in various forms (Raju, 2002). However, the internal characteristics of community participation, such as awareness creation, technical capacity building, and improving managerial skills, need a thorough understanding (Prasad, 2009). Because past experiences indicate that these factors can have a profound impact on successful introduction of renewable energy technologies through proper beneficiary identification (Maithel et al., 2015), selection of technologies (Ravindranath & Hall, 1995), changing the role of technology transfer agencies (Ravindranath & Ramakrishna, 1997), and the effective targeting of subsidies (Prasad, 2009). Hence the primary objective of this research is to understand how participatory approaches bring sustainability to energy transition strategies. The secondary objective is to identify other important factors that influence strategies, which can be combined for the successful promotion of renewable energy technologies such as biogas plants and solar lanterns among the intended user groups.

7.2 Materials and methods

7.2.1 Participator strategy for sustainable rural energy transition planning

The Ministry of New and Renewable Energy (MNRE) has various portfolio of programs that focus on integrated economic and environmental planning. The MNRE undertook an implementation program to promote renewable energy technologies in June 2011 (MNCES, 2012). As part of integrated rural development strategy, this project was designed with the objective of increasing energy supplies to rural households in a sustainable manner through the introduction of renewable energy technologies. The participation of the village community was sought at various levels and in various forms. To facilitate participation, building the capacity of the local leadership was considered essential. The ultimate goal of the participatory approach was to organize the community and build their capacities for problem solving, that is, to identify their problems, prioritize them, find, implement and manage solutions. Awareness generation was the first activity taken up before implementation. Key individuals were identified to mobilize the community. Several meetings and an interschool competition on energy and environmental issues were arranged to create awareness among the children as well as the local residents. Other activities included demonstrations of devices and their uses and the preparation of an information brochure and pamphlets on renewable energy devices to spread awareness in the community. Paintings were made on the walls of schools and houses in villages. Workshops and meetings were arranged at the subdistrict level to spread awareness in all villages.

Implementation of the designated energy intervention was carried out in a systematic manner. Training in biogas stove construction and in repair and maintenance of solar lanterns was undertaken locally involving men, women, and technicians. In the project villages, at least two people were trained for each technology and deputed for repair and maintenance. To enable communities to have ownership of the project, the communities were asked to pay some nominal amount at the start of the project. In consultation with communities, the possible contribution was decided. The community contributed a minimum of 20% of the total cost for biogas and 58% of total cost in case of solar lanterns. The remainder was matched by a governmental subsidy program. Local people were identified and trained to operate and maintain the facilities so that the local community need not depend on external agents after completion of the project. Also, necessary institutional arrangements, such as formation of village level energy committees, were made to sustain the postproject activities. This committee consisted of local masons who were provided on-the job training for construction of biogas plants and cook stoves and local technicians who were trained in the installation and repair of solar lanterns. A special committee consisting of five members was formed for ensuring overall maintenance of the infrastructures and for holding regular meetings to discuss any problems. The community was also made aware of implementation technicalities so as to allow better monitoring and vigil on quality.

7.2.2 Study areas and method

Though biogas plants and solar lanterns are being popularized in a number of states in India, their presence is prominent in a few states, such as Kerala, Andhra Pradesh, and the North-Eastern States. Within Kerala, two districts, namely, Idukki and Mallapuram, have 80 and 67 biogas plants and 4777 and 4629 lantern systems, respectively. In Andhra Pradesh, two districts, namely, Karimnagar and Khammam, are prominent for the activity. Karimnagar alone has over 16,000 units of solar lantern and 120 biogas plants. Hence Idukki, Mallapuram, and Karimnagar (Fig. 7.1) were chosen for the study. A household survey was carried out at the end of the 3-year project period. To understand the response of the recipients of the intended biogas plants and solar lanterns and the experience

Figure 7.1 Locations of the study areas in India.

they gained in acquiring the sets, 300 beneficiaries each in Idukki and Mallapuram districts and 600 in Karimnagar were selected at random for a simple questionnaire survey. Questions on the performance of technologies and the impact on the household and the community economy's were incorporated. In Kerala, grassroots-level institutions played the decisive role in identifying potential beneficiaries. In Karimnagar, much of the initiative for the popularization and distribution of the device was taken up by District Rural Development Agency (DRDA), which actually channeled the program through self-help groups (SHGs), one form of community-based organization of rural poor women. The SHGs played a crucial role in identifying the targeted beneficiaries.

The Non-conventional Energy Development Corporation of Andhra Pradesh (NEDCAP) played an institutional role in designing and operationalizing the program to popularize the lanterns among the rural poor in Karimnagar district. Apart from the beneficiaries and the SHGs, subdistrict-level development office, the DRDA, and the local cooperative bank that provided credit to the beneficiaries were the other players in executing the program. The government provided a subsidy for the devices. Additionally, to increase biomass, greening of village through plantation was planned. Informal discussions were held with all these functionaries. Similarly, the staff of the private companies that is, biogas plant designers, solar lantern distributors, and soon, who actually interacted closely and frequently with the beneficiary population for a longer period, were also interviewed to acquire greater insight into the program.

Before implementation of the program in 2004, a survey was conducted to understand the fuel use pattern in the villages. The survey also determined the willingness of individual households to pay for improved devices that were fuel efficient. Table 7.1 illustrates the fuel use pattern of the Karimnagar and

Table 7.1 Shares of different energy sources before and after the project.

Energy use (%)	Before the project (2004)			After the project (2014)		
	Idukki	Mallapuram	Karimnagar	Idukki	Mallapuram	Karimnagar
Fuel wood	83	80	75	58	55	60
Dung cakes	12	10	15	2	5	6
Petroleum gas	4	6	2	6	6	4
Kerosene	1	4	8	1	4	7
Biogas	—	—	—	33	30	23
Solar lantern (units)	—	—	—	4,777	4,629	1,600

Idukki districts before and after implementation of the energy transition program. Of all, cooking was the main energy consuming activity. The majority of households used the traditional mud stove without a chimney. Villagers extensively used fuel wood, dung cakes, kerosene, and liquefied petroleum gas for cooking. Kerosene and electricity were used for lighting purposes. However, kerosene remained the major energy source for lighting because of interruptions in the power supply.

7.3 Results and discussions

7.3.1 Participatory approaches in beneficiary identification and implementation

In Karimnagar district, Andhara Pradesh, the DRDA took the lead in introducing the biogas plants and solar lantern concept to rural people, particularly the very poor among members of SHGs, since the proposal had a subsidy component. The majority of the poor beneficiaries who were identified under the program were landless agricultural laborers, belonging to the lower social stratum and living on occasional wage income. Their economic status was so poor that they were not able to bring in even the premium amount beyond the subsidy. To help these people to acquire the threshold-level funds to participate in the program and to help them take ownership, a local cooperative bank, namely, Mulakanur Cooperative Bank, came forward with a credit line. This well-run cooperative is instrumental in giving social guidance to local SHGs and creating a social environment in which peer pressure becomes collateral for repayment. To operationalize the biogas and lantern distribution, the district officials of NEDCAP channeled the program through a grassroots-level SHG, designed to involve rural development programs. The SHGs guaranteed the loans that were raised from the cooperative bank. In the district during the last 3 years 1790 rural poor households were provided these loans by the cooperative bank at a 11% simple interest rate, repayable in four half-yearly installments. These loans were being repaid regularly. At the time of the interview, 100% of the loan obligations were being honored.

In Kerala the identification of beneficiaries was carried out through local institutions operating at the grassroots level. These institutions include village-level committees known as Ayalkoottams, beneficiary committees, and neighborhood groups. These institutions have become quite active and powerful since 1997–98, following the substantial fund devolution from the state resources. By and large, the beneficiary selection process was transparent. The process was vetted and put to debate on at least three or four occasions before the final decision was made. The identified list of candidates was finally acceptable to all. Such a broad concern is possible only as a result of participatory planning. The process started with the identification of needs in the subvillage committees, which were constituted in wards. This was followed by secondary data collection, such as preparing a socioeconomic profile of the intended beneficiaries. The output of these exercises were integrated into a documented report, which was later discussed in a village-level meeting of around 250–300 people known as a development seminar. The gathering also assessed the needs of individuals, prioritized them, and prepared the list of beneficiaries. After the list was approved by the village committee, it was sent to a block-level expert committee consisting of retired and serving officials for vetting the project. Once this was done, the list was sent to the district committee that was responsible for implementing the program. The entire process stressed principles associated with good governance, such as transparency, the right to information, a social audit, people's participation, and equity.

7.3.2 Performance of renewable energy technologies

The household survey suggests that 93% of biogas plants and 72% of solar lantern sets in Idukki, 97% and 82% in Mallapuram, and 94% and 80% in Karimnagar, respectively, were working satisfactorily. The details of the questionnaire and interview survey are reported elsewhere (NIRD, 2014). The high level of technology performance suggests simplicity of the technology, adequate training, and of course the service backup. The biogas plants and solar lanterns distribution in Mallapuram was taken up some 3 years ago, whereas in Idukki their distribution was taken up after about 2 years. Compared to the Kerala experience, the solar lantern project in Karimnagar is new, yet a large proposition of lanterns in this district was found to be working in the study area. In Mallapuram and Karimnagar 97% were working while as in Idukki this was 94%. Among the lanterns, those that were a year old or more were examined for their performance. About 90% of them were found to be performing well. The enquiries suggest that in Mallapuram, 67% of lantern nonperformance incidents were due to failure of the battery. Battery failure accounted for 44% of the device failures in Idukki and 45% in Karimnagar. Negligence in following the suggested schedule for battery recharge and upkeep was another important reason leading to the nonworking of the devices. In Karimnagar and Idukki, negligence was responsible for about 40% of the failure cases. It may be noted that the nonobservance of the suggested schedule minimal in of Mallapuram. This suggests that the technology users improve their conviction as well as confidence in the device if participatory planning and capacity building program precedes implementation of the energy transition programs.

7.3.3 Impact of the rural energy transition program on community livelihood

The energy transition program through community participation has enabled the rural poor living in undeveloped areas to have environmentally friendly lighting facilities. It enhanced their local mobility and helped their children to increase study hours. In Karimnagar the lanterns helped the households save 2–3 L of kerosene a month, which in monetary terms works out to about Rs. 25. The program also helped women to realize that if they organize themselves into strong and active groups, they can gain access to development programs. Biogas plants and solar lanterns became one of the developmental issues in the internal meetings of the SHGs and in meetings with the cooperative bank officials, service agents, and DRDA officials. These technologies also figured very prominently in the credit review meetings of SHGs. For several SHG members the popularization and adoption of biogas and solar lanterns turned out to be a good reason to improve community participation in other development activities. The SHGs in the project area were well consolidated (Thomas & Lal, 2007), because of which they responded well to the idea from DRDA. For a few, the solar lantern proposal turned out to be a lead issue, along with work, shelter, health, education, and credit. For 6 months it became a rallying point.

The appraisal of costs involved in operating the project suggests no or little cost on the part of either NEDCAP or the DRDA, the project implementing agencies. In all these departments the staff time allocation for the project was very limited. SHG meetings and their organization turned out to be the focal point of interaction, which was carried out with minimal or no cost to the state. In Kerala the lanterns enhanced the villagers' mobility. The project helped in improving the energy conservation of more than 75% of households (NIRD, 2014). It is estimated that every year almost 32 tons of fuel wood and about 10,000 L of kerosene will be saved by a population of 75 households (Purushotam, 2004). In other words, it may be concluded that fuel wood saving may help to increase the productivity of the available trees. Maintenance of growth of such available trees will eventually reduce soil erosion and help to mitigate global climate problems. Potable water is also available in the Mallapuram village through the installation of a solar pump. Installation of 13 biogas plants has provided good quality of manure and sufficient cooking gas for daily cooking for six to eight family members. The community-based process for implementation of clean energy intervention has also led to other environmental benefits, such as reduction of the indoor pollution and reduction in carbon dioxide emissions. The improved fuel-efficient devices, such as biogas plants, improved stoves, solar and lanterns, are reducing indoor pollution levels to a minimum (Purushotam, 2004). Also, dissemination of renewable energy technologies helped in reducing the drudgery of women and children through using improved cooking and lighting devices, namely, biogas and solar lantern. As of now they do not have to collect fuel wood from long distances.

7.3.4 Community participation and rationale for subsidies in rural energy transition programs

The biogas plants and solar lanterns are the most appropriate device to provide cooking and lighting, respectively, to the inhabitants living in interior areas not connected by the central electricity grid. Nevertheless, the technologies have become popular among the target beneficiaries in a short time because at such the can obtain lighting technology at a low investment. An average solar lantern in 1999–2000 was priced at Rs. 3100, and the price dropped to Rs. 2900 in the next 2 years and further to Rs. 2700 by 2002, owing to a reduction in import duties as well as tax reforms with regard to imported photovoltaic cells. The MNES used to provide a subsidy of Rs. 1500 on each unit until 1999–2000, which was later reduced to Rs 1300 by 2001–02. With such assistance the lantern was available for Rs. 1700 to the beneficiary. Rural people could really benefit from this measure. In Karimnagar, even the very poor with the strength of SHG institutions and cooperative banks were able to benefit. The subsidy was withdrawn in 2002, meaning that one has to buy the device at Rs. 2700, which is unthinkable for the typically poor. Withdrawal of the subsidy was based on the apprehension that it was not reaching the targeted poor (Purushotam, 2004), and it is sought to be justified by rerouting a small part of the subsidy through the solar lantern manufacturers by lowering the customs duty on the imported solar photovoltaic cells and providing institutional financing with liberalized repayment schedules and lower interest rates. This measure has brought the cost of production from Rs. 3100 to Rs. 2700 in 4 years. Because of the marginal fall in the device prices, its market is believed to have grown slightly (MNCES, 2012). However, this marginal price drop could not attract the rural poorer sections. The withdrawal of the subsidy because of the supposition that it does not reach the poor was not based on a true assessment of the situation. When the program is channeled through the involvement of people's institutions, such as village committees in Kerala and SHGs in Andhra Pradesh, the subsidy can indeed reach the deserving poor, and such assistance could enhance their threshold-level resource position to receive the benefits of the program.

The introduction of these technologies has not only provided self-employment or microenterprise development, as detailed in Table 7.2, but also motivated a few among the middle and higher income groups in rural areas to patronize the product. In Karimnagar, where the tariff for conventional electricity is very high and the quality of the power supply is poor, owing to power cuts and voltage fluctuations, the solar lanterns are gaining popularity among the nontarget groups, even at market

Table 7.2 Economic performance indicators of enterprises servicing solar lanterns.

Indicator	Idukki	Mallapuram	Karimnagar
Number of solar lantern servicing units	12	16	10
Number of sample respondent units	7	7	9
Average total investment (rupees)	32,800	40,500	22,115
Average monthly gross income (rupees)	7,850	7,648	4,414
Average monthly operational expenditure (rupees)	1,650	2,110	1,200
Average monthly income from solar lantern servicing (rupees).	1,984	2,160	900
Average monthly net income (rupees)	6,200	5,538	3,214
Percentage solar servicing income in total income derived from other agricultural related activities (%).	25.27	28.24	20.38
Growth of income of the solar service agents during the last 3 years (average, %)	8	6	16
Average number of workers employed by a unit (in addition to the self- employed)	1.4	1.7	1.1
Received institutional loan assistance	2	1	—

prices. Thus dual pricing of the product is not only desirable but also feasible. The observations from Karimnagar, Idukki and Mallapuram suggest that the government's macro-policy regarding the withdrawal of subsidy on solar lanterns and rerouting it through the manufacturers needs a review and reconsideration. Rural energy technologies such as biogas and solar lanterns can be taken to the target population only through rational and well-focused public policies, meaningful involvement of the community in the beneficiary selection, and operationalization of the technology diffusion or popularization schemes.

7.4 Conclusion

Rural energy transition strategies depend on a number of factors, including the technology (transmitter, receiver etc.), the technology user population, the institutions that are taking responsibility for technology popularization, national policies, and, most important, community participation. Successful dissemination and popularization of renewable energy technologies such as biogas and solar lanterns meant for the households in undeveloped rural areas are possible only if the target beneficiaries are included in the decision-making process before, during, and after the implementation of the projects. The public policy regarding renewable energy technology transfer should also encourage community initiative and an entrepreneurial spirit. The biogas and solar lantern program in the study area suggests that success in technology popularization greatly depends on the involvement of the people's institutions, such as village committees and neighborhood committees, in beneficiary selection. In fact, the beneficiary selection process should be transparent, democratic, and inclusive in any integrated rural development project. If development financial institutions can work out microcredit lines, the technology diffusion and transfer could be made more effective. Furthermore, popularization of such low-end technologies in areas with underdeveloped infrastructure crucially depends on the growth of microenterprises to attend to the service needs of the technology. This case study, carried out by interviewing beneficiaries of a government subsidy program in three rural districts in India, suggests that effective implementation of energy transition strategies and their sustainability can be facilitated by meaningful participation of the community, which also facilitates defining roles clearly for other stakeholders. Importantly, a participatory approach ensures that rural energy planning is also in line with the developmental needs of the community.

References

Haq, M. U. (1976). *The poverty curtain: Choices for the third world.* New York: Columbia University Press.

John, M. S., & Chathukulam, J. (2001). Social capital formation and institutional growth following participatory planning in Kerala: A critical appraisal. In P. Purshotam, & V. V. Reddy (Eds.), *Emerging institutions for decentralized rural development* (pp. 458–464). Hyderabad: National Institute for Rural Development.

Maithel, S., Dutta, S., Prasad, R., Kumar, A., Sing, P. B., Pal, R. C., … Sharma, S. P. (2015). *Fuel substitution in the rural sector.* New Delhi: The Energy Research Institute, Report no. 95/RE/62. p. 138.

MNCES. (2012). *Annual report, ministry of non-conventional energy sources.* Government of India.

NIRD. (2014). *Annual report 2003–2004, national institute for rural development.* India: Hyderabad.

Norman, C. (1978). *Soft technologies, hard choices.* Washington DC, USA: World Watch Institute.

Prasad, R. (2009). Community participation in the development of an improved stove in cold region of North India. *Boiling Point, 42,* 30–32.

Purushotam. (2004). *Report on promotion of renewable energy technologies in backward districts of Andhra Pradesh and Kerala.* Hyderabad, India: National Institute for Rural Development.

Raju, S. K. (2002). Issues in transfer of rural technologies. In Purushotam (Ed.), *Rural technology for poverty alleviation* (pp. 126–138). Hyderabad: NIRD.

Ravindranath, N. H., & Hall, D. O. (1995). *Biomass energy and environment: A developing country perspective from India.* Oxford University Press.

Ravindranath, N. H., & Ramakrishna, J. (1997). Energy options for cooking in India. *Energy Policy, 25,* 63–75.

Reddy, A. K. N., & KrishnaPrasad, K. (1977). *Technological alternatives and Indian energy crisis. Economic and Political Weekly* (pp. 27–36).

Thomas, I., & Lal, H. (2007). Planning for empowerment: People's campaign for decentralized planning in Kerala. *Economic and Political Weekly,* 53–58.

8

Energy, peace, and nation building in South Sudan

David Mozersky[1] and Daniel M. Kammen[2,3]

[1]Energy Peace Partners, San Francisco Bay Area, United States [2]Energy and Resources Group, University of California, Berkeley, CA, United States [3]Goldman School of Public Policy, University of California, Berkeley, CA, United States

Chapter outline

8.1 Introduction and history

Things have not gone according to plan for South Sudan. After gaining independence in 2011 amid great hope and international celebration, South Sudan has seen stagnating economic and political progress exacerbated by renewed civil war erupting in late 2013. International donor governments continue to support the billion-dollar-per-year humanitarian operations in the country, but humanitarian funding is under stress around the globe, and the South Sudan crisis shows no signs of abating. Most donors do not want to support the current government directly because they fear that funds could be diverted

Energy Policy for Peace. DOI: https://doi.org/10.1016/B978-0-12-817350-3.00007-9

to military efforts, which are likely to target civilians. Donors also recognize that short-term humanitarian programming alone will not end the crisis.

What can be done to generate new opportunities and momentum for a more peaceful future when the outlook is so bleak? This chapter argues that a system-wide donor-driven transition to renewable energy, specifically solar power, to support humanitarian programming is a viable way forward both now and over the longer term. Although such a transition alone will not end the conflict, it offers donors a more strategic alternative to the current practice. In the near term, a pivot to renewable energy will offer significant cost savings in a nation where electricity generation is one of the highest recurring costs in humanitarian budgets. Over the longer term, this approach will create long-lasting, reliable energy infrastructure and building blocks for peace and development in the least electrified country in the world. At a minimum, a shift to greater domestic reliance on renewable energy will help to decouple economic growth from the geopolitics of oil and gas at a time when the growing impacts of climate change highlight South Sudan's vulnerabilities (Stalon & Choudhary, 2017).

Even before the outbreak of conflict in 2013, South Sudan had the lowest electricity consumption per capita in the world and ranked near the bottom in many global development indicators (IEA, 2016).[1] The modest progress that was achieved during the peaceful years between 2005 and 2013 has largely been undone by the conflict since then, and much of the infrastructure, energy and otherwise, has been destroyed or looted. The electricity generation that does exist across the country is intermittent and comes almost entirely from imported diesel for generators. The lack of electrification affects all sectors of society, with government offices, hospitals, and even the national parliament coping with regular blackouts (Alstone, Gershenson, & Kammen, 2015). As the economy has collapsed and security has worsened, diesel supply lines have become less reliable and more expensive. Rampant inflation has exacerbated the problem, the national consumer price index increasing more than 2100% between December 2015 and July 2017, making fuel and other staple products much more expensive (World Bank, 2016) (Fig. 8.1).

[1]According to the International Energy Agency (IEA), South Sudan averaged only 39 kWh of electricity consumed per capita for the entire year of 2014. This put South Sudan alongside Haiti at the bottom of the national rankings (IEA, 2016). By comparison, Ethiopia averaged 70 kWh, Kenya 171 kWh, Mexico 2169 kWh, and the United States 12,962 kWh. By 2020 the expansion of supply amounted to only 49 kWh per capita.

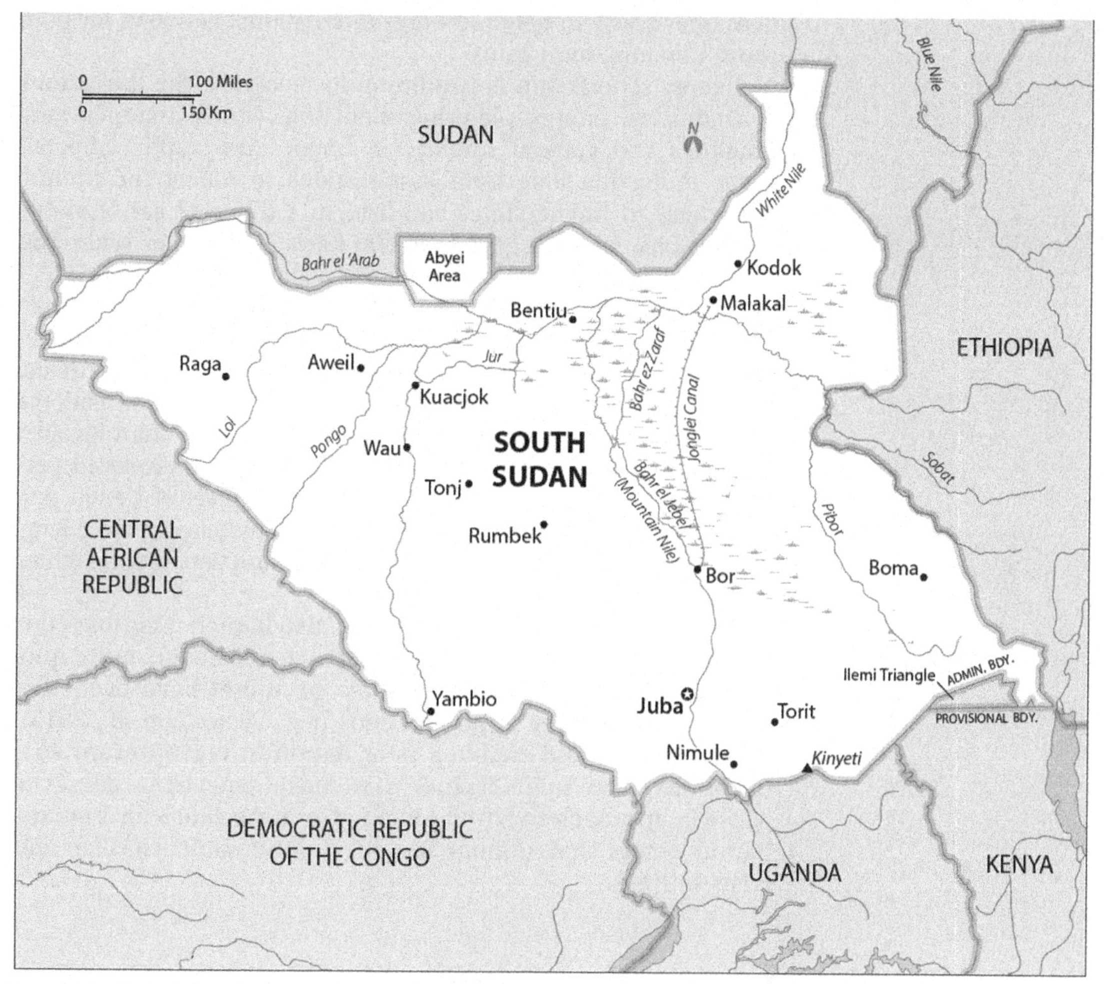

Figure 8.1 Map of South Sudan with key energy assets. From Lucidity Information Design, LLC.

Renewable energy offers tangible and immediate benefits that bear out over the long term. The cost of solar power in particular has dropped dramatically in recent years, and solar now is both a cheaper and a more consistent power source than alternatives in South Sudan. Solar panels can be easily scaled and can last for more than 20 years. Donor support for a solar push could help to create new jobs and enhance the sustainability of local capacity. Renewable energy alone cannot end the conflict, of course, but it does offer a way to better leverage international aid flows for both short- and long-term gains.

Energy scholarship is beginning to recognize the limitations of traditional energy planning modeling, which assumes long timelines and general stability in fragile and conflict-affected states. Adjusting long-term assumptions to reflect the unique challenges in fragile states can lead to a different set of viable conclusions, including greater emphasis on smaller-scale and renewable energy systems (Bazilian & Chattopadhyay, 2015).

Three immediate opportunities are anchored in humanitarian programming in which solar investment by donors can yield benefits. One is onsite energy generation in individual Nongovernmental or government compounds. The second is in neutral national health institutions, such as hospitals and clinics. The third involves the protection of civilians (POC) sites for internally displaced persons (IDPs) outside the destroyed regional capitals of Bentiu and Malakal; in these locations, solar systems would also create energy assets that could transition to support reconstruction and the return of IDPs to the cities in years to come.

A shift to renewable energy could also launch a far less contentious resource base than the current fossil fuel status quo. The negative impacts of a reliance on diesel have been discussed in a number of publications (e.g., Alstone et al., 2015). A solar pivot could enable a poor nation to move toward sustainability. The same factors that make renewable energy a win-win approach in South Sudan also make sense for humanitarian actors and donors in other comparable conflict and crisis settings.

8.2 Energy development in South Sudan

The Southern Sudanese government received close to $13 billion in revenue between 2005 and 2011, and oil accounted for more than 98% of the total intake (GRSS, 2011). Over the same period, the international community spent approximately US$5 billion in development aid in South Sudan and invested another US$5.76 billion in the UN Mission in Sudan (UNMIS), which covered both Southern Sudan and parts of northern Sudan (Global Humanitarian Assistance, 2011).

Despite the money flowing into Southern Sudan, little energy-related infrastructure was built. Government electrification plans centered on the construction of several Chinese-led mega-dams along the White Nile as part of a long-term plan to build out a national grid. Ground was never broken on these projects. In the meantime, international donors, UN agencies, nongovernmental organizations (NGOs), the regional government, and private sector actors spent hundreds of millions of dollars on generators and diesel fuel to power both aid and development efforts. Diesel was widely available from Sudan, which ran its own domestic oil refinery. The Juba government was able to pay for this fuel using Sudanese currency that it received for its oil sales from the central government in Khartoum. In Juba and other southern cities, networks of diesel generators were installed or expanded to power new city grids. A Norwegian-supported hydroelectric project at the Fula rapids, near the border with Uganda, was developed to provide power to Juba, but progress was slow.

Several other projects were launched to build domestic oil refineries, but these too moved slowly and had ground to a halt by 2014 because of the economic crisis and the new civil war, and none was ever completed. In March 2017 the government announced its latest plan to build a new oil refinery in South Sudan to be operational by mid-2017.

When South Sudan became independent in 2011, not all the terms of separation from Sudan had been agreed. The most notable outstanding issue was oil, a major economic lifeline for both Juba and Khartoum. Oil was a main economic driver for Sudan and the largest source of government revenue and foreign currency, but approximately 70% of the oil produced came from oil wells in South Sudan. As a newly independent country, South Sudan no longer had to submit to the fifty-fifty revenue-sharing arrangement with Khartoum that had existed during the 2005–11 interim period, but Juba still relied on the oil pipeline that ran through Sudan to get South Sudan's oil to market via the Red Sea. Sudan hoped to charge high transit fees for the use of the pipeline to make up for its lost oil revenue; South Sudan offered a much lower transit fee, hoping to leverage a broader transitional payment to facilitate agreements on other outstanding issues. As the negotiations continued, the oil continued to flow with no formal agreement in place.

In January 2012, after not receiving payment for the first 5 months after South Sudan's independence, and as the economic impact of reduced oil revenue began to bite, the Sudan government took matters into its own hands and began unilaterally offloading southern oil directly from the pipeline at Port Sudan as payment in kind and reselling it internationally. The South Sudanese government responded by shutting down its entire oil production rather than having the oil taken by Khartoum. The popular decision was initially cheered in the streets of South Sudan, where the population viewed it as a symbol of the young nation's independence from its old oppressor. Harsh reality quickly set in, however, as the economy ground to a halt. Southern officials scrambled around the region trying to find support for building an alternative pipeline to the sea through Kenya or Ethiopia and Djibouti, but no short-term solution could be found. Meanwhile, government foreign currency reserves began to dry up, and the economic situation worsened.

Electrification suffered immediately. The diesel generators powering Juba broke down in 2012, and funding to maintain an adequate flow of diesel or to repair the generators ran short. The Government of South Sudan needed to use its foreign currency reserves to purchase fuel imports, and dollars were suddenly scarce. South Sudan's economy began slowly but steadily to collapse. Plans to build domestic oil refineries stalled indefinitely. The negotiations over pipeline access with Sudan continued, but the balance of power had now shifted significantly in Khartoum's favor. An agreement was eventually reached, and pipeline access and southern oil production resumed in the first half of 2013. In mid-December 2013, violence broke out in Juba after a political disagreement between President Salva Kiir and former vice-president Riek Machar. Initially concentrated in Greater Upper Nile, where the country's oilfields are located, the conflict eventually spread across much of the country.

8.3 Impact of renewed conflict and economic collapse

Before war broke out in 2013, South Sudan had just 22 MW of operational installed electricity generation capacity out of 30 MW coming exclusively from diesel and heavy-fuel generators, most of them located in a handful of cities (Altai Consulting, 2014). For comparison purposes, 22 MW is equivalent to the electricity required to power approximately 3600 homes in the United States. As of August 2017, operational generation capacity was almost certainly considerably lower, given the extent to which power generation had been disrupted or destroyed in fighting since 2013.

The impact of the precipitous drop in southern oil production caused by the closure of the Unity State oilfields was exacerbated by the dramatic drop in global oil prices. The Dar blend from South Sudan's still-operating Upper Nile fields trades at an additional discount on the international market because of its lower quality. Furthermore, the Juba government must pay other costs (such as profit-sharing and processing fees) to the oil companies and the pipeline transit fees to Khartoum. Oil production remains the government's primary source of revenue but has brought in very little since 2014. Inflation and divergence between the official and black market exchange rates have both spiraled, making goods, including diesel, far more expensive. The dramatic increase in the price of gasoline and diesel following the devaluation of currency led to a noticeable reduction in demand, with fewer people being able to afford them (Hoth Mai, Mayai, & Tiltmamer, 2016). One effect of this trend is that internationally funded programming now offers the best entry point for renewable energy in South Sudan.

Despite the economic crisis, the government continues to import and heavily subsidize (by as much as 80%) the domestic sale of gasoline and diesel, which are supposed to be sold to the public through gasoline stations at a discounted rate. Nonetheless, the black market for both gasoline and diesel has remained active, and the price for a liter of fuel not only varies widely across the country but is also well above the official rate in every location. The black market for fuel is visible on street corners in Juba, where lines of cars waiting at gasoline stations stretch down streets and around blocks. Traders make a fortune buying up government-subsidized petrol and then reselling it on the black market for up to five times the price they paid (Akec, 2016). The South Sudanese government may have spent as much as 20% of its entire 2017 budget on fuel subsidies, but the benefits were being captured by a group of traders rather than being distributed among the population at large (Akec, 2016). The deepening fuel crisis, which saw a liter of diesel fetching as much as SSP400 in Juba in late July 2017, eventually prompted government promises to end the fuel subsidies and crack down on black market trade (Radio Tamazuj, 2017). Reports in early December 2017 indicated that the government had quietly ended its fuel subsidies because it did not have the resources to back them (Sudan Tribune, 2017b).

8.4 The opportunity for renewable energy

As a global resource, renewable energy has come of age, with the sector seeing its largest annual increase in capacity in 2015. Wind power and solar photovoltaics (PV), the most popular form of solar power—seen in most solar panels today—accounted for more than three-quarters of new energy installations globally in 2015, followed by hydropower. The world now adds more renewable power capacity annually than it adds (net) capacity from all fossil fuels combined. By the end of 2015, installed renewable capacity was enough to supply almost one-quarter of global electricity (REN21, 2016).

Although advances in all areas of renewable energy supply are relevant, the evolution of the solar energy sector, in particular of solar PV, has potentially the most immediate importance for South Sudan. Solar PV is notable for its ease of installation and operation at all scales, from solar lanterns to rooftop systems for houses the fast-growing area of minigrids, and large utility-scale systems. A particularly important aspect of the evolution of solar PV energy for South Sudan is its scalability. At the bottom end of the scale, the off-grid, pay-as-you-go market has exploded in East Africa; a range of companies now offer systems in the 0- to 150-W peak range that are attractively priced, often require no down payment, and can run appliances such as radios, televisions, and refrigerator-freezers. Solar power also plays a key role in minigrids, which can be 1 MW or more in size and can be powered by solar-plus-storage systems (i.e., solar panels plus batteries, which store energy for use when the sun is not shining); by solar-hybrid systems in which solar energy is combined with diesel, wind, hydro, or other technologies to provide energy for communities or businesses; or by a combination of the two. This rapidly expanding sector brings many of the benefits of large utility-scale grids without the larger infrastructure costs (and vulnerabilities) that war-torn regions such as South Sudan will struggle to overcome for the foreseeable future.

Systems at each of these scales could play valuable roles in the "green pivot" that is being proposed for South Sudan. Because large, utility-scale solar projects are not feasible until the country is more stable, the immediate focus is on smaller-scale systems. When stability is achieved, smaller-scale systems could become a major component of a vibrant domestic and export clean energy economy.

8.4.1 Advantages of solar power for South Sudan

Renewable energy, particularly solar power, has the potential to be transformative in South Sudan for several reasons. First, compared with energy produced by diesel generators, renewable energy is cheaper, cleaner, and longer lasting; whereas a diesel generator requires new diesel to burn and the generator itself must be replaced every few years, a solar panel can reliably produce electricity for up to 25 years. Second, South Sudan has ample sunshine with strong solar power potential (high solar irradiance). Third, as the conflict drags on, internationally funded humanitarian programming can launch an expansion in solar power generation, offering short-term energy benefits and cost savings while building an enduring infrastructure that will outlive the conflict and contribute to peace and stability over time.

Despite donor fatigue, large-scale international humanitarian aid funding is likely to continue in South Sudan for the foreseeable future. Currently, international donor governments cover the high energy costs for the purchase of generators and diesel for individual NGO compounds and programs, UN agencies and peacekeeping bases, and POC camps. These costs can account for a significant proportion of program budgets, particularly for activities outside Juba, and must be paid year after year, leaving nothing to show for the expense. Given the high price of diesel in South Sudan, renewable energy systems could quickly pay for themselves, potentially saving millions of dollars in the years to come.

Unfortunately, existing diesel usage and energy figures are not readily available from international donors or humanitarian NGOs. Some limited information about fuel usage by the UN Mission in South Sudan (UNMISS, the successor mission to UNMIS, which ended in July 2011) is publicly available, but the information is not broken down explicitly between energy and transportation usage. A late 2014 audit report shows that the peacekeeping mission signed a fuel contract in March 2014 that was capped at US$325 million over the 3-year life of the contract. During the previous 3 years, from July 2011 to May 2014, UNMISS had received 68.7 million L of diesel fuel, which was used for vehicle transport and to power generators (UN, 2014). As of June 2015, UNMISS owned and operated 195 diesel generators in 22 locations across the country (UN, 2015).

The need for international organizations and NGOs to reduce costs is increasing as the gap between humanitarian funding needs and funding pledges widens. In August 2016 the then US secretary of state John Kerry threatened to halt US government aid to South Sudan if the warring leaders did not do more to end the conflict (Stevens, 2016). The US government is by far the largest donor to South Sudan, having spent more than US$2.15 billion between 2014 and March 2017.[2] The cuts in US foreign aid proposed by the Trump administration may well affected US funding in South Sudan, among others.

US-funded humanitarian operations, as well as those of the UN and other international actors in the country, could make better use of their diminishing resources. A few NGOs and individual agencies in South Sudan use solar power successfully.

[2]This figure includes funding for South Sudanese refugees in neighboring countries (USAID, 2017).

Internews, for example, which supports local radio stations across the country, struggled for years with expensive generator maintenance and inconsistent diesel supply lines before transitioning its more remote radio stations to 100% solar-plus-storage power. Following the successful transition to solar energy at its station in Turalei in Bahr el Ghazal, Internews worked with other community radio stations to help them make a similar transition and improve their energy supply.[3]

Such success stories should not obscure the significant challenges facing widespread solar adoption in South Sudan. Looting and destruction of humanitarian agency property during the civil war have targeted solar panels (and diesel generators) in population centers and rural humanitarian outposts. This in part explains international donor hesitancy to invest in solar systems (which are more expensive to purchase initially than diesel generators). Other impediments are the limited capacity in terms of trained personnel and the economic and security environment of the solar sector. These challenges can be at least partially mitigated, however. Security risks can be minimized with the strategic placement of solar systems. Local capacity can be built and supported through donor-led investment in training and capacity building to help ensure that the South Sudanese benefit from such a transition and that solar systems are adequately maintained.

8.4.2 A donor-led transition to renewable energy

The advantages of a system-wide, donor-led pivot to renewable energy for the humanitarian sector in South Sudan are many. The transition would pay for itself and begin to generate cost savings within 2–5 years. It would create desperately needed energy infrastructure while supporting ongoing humanitarian operations. These long-lasting, clean energy assets could support future reconstruction and health, education, and social service delivery and would carry the additional benefit of being in place and, likely, of having already been paid for. The pivot would also create new entry points for conflict resolution and peace building. For example, the transition from a solar system that serves a humanitarian program to one that supplies a local institution (e.g., a local utility or hospital) would generate opportunities for cooperation between communities that have been driven apart by the civil war to determine issues such as the placement, oversight, management, and maintenance of the solar system. Such a shift would create

[3]Interview with Internews officials, Juba, May 2016. See also Nikolov (2017).

physical assets for peace that could both support physical reconstruction efforts and provide new hooks for conflict resolution.

Donor-level support for this kind of transition is important for two reasons. First and most obviously, donors provide the funding to sustain current humanitarian operations, including existing diesel energy budgets. Donors are thus the primary option for funding a system-wide transition to solar power. Other conflict-affected countries, Syria and Yemen among them, have seen piecemeal efforts to use solar power, but nowhere has a concerted effort to accomplish a large-scale transition been undertaken (e.g., Mansoor, 2016). Second, if solar energy infrastructure is to transition to local ownership or to benefit local communities in the medium to long term, it needs to be managed and maintained by actors that are likely to remain in South Sudan over a similar period. Unlike individual humanitarian organizations, donor governments generally operate with a national or regional lens and a longer-term perspective. Their mandates and missions will evolve to support development and reconstruction when the situation allows for it. Donor-level support for a shift to solar power could thus ensure continuity and help to create and anchor a framework for managing the next phase of such a program.

Making a system-wide transition would break new ground for such a comprehensive use of renewable energy in other crisis settings. Although some humanitarian actors are beginning to explore the potential for renewable energy, it has yet to become a mainstream practice. The Moving Energy Initiative, funded by the UK Department for International Development, is an ambitious multi-organization consortium that is exploring how to integrate renewable energy into refugee camp settings, focusing first on Kenya, Jordan, and Burkina Faso. The United Nations has started to prioritize renewable energy in its peacekeeping operations under its Greening the Blue Helmets program, but these efforts are still early stage, with few successful examples. Leveraging this UN transition to also support economic, local electrification, or peace benefits remains more hope than reality at this time.

In South Sudan the transition to solar energy is possible in at least three distinct types of humanitarian programs or operations in South Sudan: individual NGO compounds, hospitals, and POC sites.

8.4.3 Individual nongovernmental organization compound solar systems

Practically all electricity in South Sudan is generated by diesel generators. Internationally funded humanitarian NGOs and UN agencies run their own generators to power their compounds and programs, as do government ministries and offices, even in the main ministerial compound in Juba. Relatively small-scale solar systems could be installed in individual NGO compounds and positioned on the ground, on rooftops, or on shipping containers. As has been noted, solar systems are often coupled with backup battery storage, and hybrid systems include both a solar component and a diesel generator. Unlike some peace dividends that have proven elusive in postconflict settings, a solar system would yield immediate, observable benefits by reducing diesel costs and reliance on inconsistent supply chains (Wade, 2017). More important, once in place, these systems could easily be extended and connected to neighboring compounds or houses to create local minigrids, which could eventually be connected to a larger city grid.

Donors may have to create incentives for their humanitarian grantees to make such a switch. Most NGOs are not in the business of creating access to energy, so a strong push from donors and their continued involvement will be necessary to build out minigrids anchored around internationally funded, NGO compound–based solar systems.

8.4.4 Hospitals and health infrastructure

South Sudan's health infrastructure was weak before the civil war and has declined further since fighting resumed in 2013. The main hospitals in the regional capitals of Malakal and Bentiu have been targeted during the fighting. On the long list of needs for most hospitals and clinics, energy supplies are near the top. For example, Juba Teaching Hospital, the country's only referral hospital and the main civilian hospital in the capital, has struggled through regular and extended power outages since the Juba grid collapsed in 2012. The hospital depends on inconsistent diesel stipends provided by the Ministry of Health. During the two outbreaks of fighting in Juba in December 2013 and July 2016, the hospital was flooded with patients and struggled with blackouts because the fighting shut down diesel supply lines. Even when diesel is available, the hospital is forced to ration its fuel supplies by shutting down its generators overnight, and sometimes it must resort to purchasing additional diesel on the black market. Similar constraints have affected hospitals and health clinics across the country.

Diesel-powered generators are a dirtier form of energy than renewable energy and may cause even more pollution. A report published in September 2016 found that diesel sold in Africa was the dirtiest in the world; its average sulfur content is at least four times higher than that found in any other region and 200 times higher than European levels. This variance is due primarily to lax national regulations and unscrupulous petroleum actors. It is associated with a range of negative health impacts, including increases in respiratory and cardiovascular diseases (Gueniat, Harjono, Missbach, & Viredaz, 2016). The report did not include South Sudan, but anecdotal evidence suggests that much of the diesel that is being sold in South Sudan is of particularly low quality and contains many impurities, causing significant problems for running generators and leading to lower productivity and higher maintenance costs.

Support for the health sector, and service delivery more generally, is an attractive choice for international donor funding. Investment in renewable energy for hospitals and clinics could significantly improve healthcare capacity simply by providing consistent and reliable energy. Hospitals would also benefit from the same economic savings that NGO compounds with solar systems would enjoy as a part of a local minigrid, thereby increasing energy access for the surrounding communities.

South Sudan's international partners face a dilemma: whether to support or circumvent the state. Most do not want to provide budgetary support to the government because they believe that funds may be diverted to military efforts, which are likely to target civilians. At the same time, international partners appear to believe in state continuity. When Juba was at risk of falling, international partners supported the existing order, perhaps believing it to be the only alternative to state fragmentation or collapse. Many donors are managing this dilemma by switching their assistance to the humanitarian sector and disbursing aid through international organizations and local NGO partners. The problem with this approach is that the South Sudanese state is withering into the space created by its own tribalized security forces. This process is further militarizing society and undermining civil spaces everywhere. But civil spaces still exist. In many churches and mosques, schools and universities, and hospitals and health centers, civilians are trying to maintain shared spaces and protect their shared identities.

Supporting spaces that resist the militarization and tribalization of society offers at least three potential peace-building gains. First, it is a way of preventing a drift toward genocide or further violent deterioration of interethnic relations. Second, it supports the survival of nonmilitarized social and political values and possibilities, crucial for any future peace agreement to take root. Third, it supports a decentralized version of development, which could help mitigate the centralizing tendencies of the security state.

8.4.5 Protection of civilians camps in Malakal and Bentiu

Perhaps the most compelling case for investing in renewable energy systems is in the large POC sites, that is, the IDP camps that are housed within UN peacekeeping bases and are home to roughly 200,000 civilians who have fled the violence in Juba, Bor, Bentiu, Malakal, Wau, and Melut. The Bentiu and Malakal POC sites are two of the largest camps in the country, housing approximately 110,000 and 35,000 IDPs, respectively. Both camps are situated on the outskirts of regional capitals that were destroyed during the civil war. These displaced populations are supported by large humanitarian operations powered entirely by diesel; the annual cost of powering each camp's humanitarian operations is approximately US$1 million. Malakal relied on air shipments of diesel fuel (and everything else) because of insecure road and river access until 2020, which the International Organization for Migration (IOM), which runs the Malakal Humanitarian Hub, successfully implemented a 700 kW private-sector led solar project to power the Hub. Bentiu is accessible by road for part of each year, which makes for slightly lower but still expensive diesel costs.

Because of the high cost of diesel, renewable energy systems in these camps would offer rapid cost savings for humanitarian operations; the expenses that are involved in the stalling renewable systems in Malakal's new solar project has been cost-effective for both IOM and the private sector partner; the outlay for Bentiu would be repaid within 3–4 years. Large-scale civilian returns to Bentiu and Malakal are unlikely in the near term, given the ongoing conflict and continued local and national tensions, but the civilian populations will probably opt to return to their homes when adequate peace, security, and stability have been restored. Investment in renewable energy infrastructure today can help to build the power plants of tomorrow for these cities. Transitioning the larger UNMISS energy footprints to solar would further expand this clean energy infrastructure. Continuing with the status quo of diesel power will mean that when the situation improves enough for displaced civilians to return to their homes, humanitarian actors will either pack up and leave or transition to new locations, and donors will begin to think about supporting reconstruction in the cities. Creating new solar systems for humanitarian operations now would create an energy infrastructure that will be able to transition from camp to city to support reconstruction and returns in Bentiu and Malakal in the years to come.

In addition to being subject to broader conflict dynamics, both Bentiu and Malakal struggle with local conflict drivers that must be resolved for peace to take hold. Solar systems with battery storage would create pro-peace assets that could both serve as entry points for promoting local cooperation and conflict prevention and form the core of a new local electric grid. If the security of the solar systems is a concern, the systems could remain physically located within UN bases—such as the solar system in Malakal Humnitarian Hub—and be connected to the cities by wire. The near- and long-term advantages of solar systems for Bentiu and Malakal could also be enjoyed at other locations in South Sudan, including in the IDP camp in Wau and in refugee camps hosting Sudanese civilians in Maban and Yida.

Donor funding generally operates in distinct categories: South Sudan's aid is primarily humanitarian crisis funding, which usually operates on short-term funding cycles and is distinct from reconstruction or development funding. Renewable energy can bridge these categories. It offers an immediate cost-saving strategy while creating building blocks for future peace. The POC projects in particular have to navigate tricky political currents. UNMISS has at times been unhappy that its bases are still being used to host and protect nearly 200,000 civilians. It also remains sensitive to infrastructure projects that carry any hint of "permanence" in the camps.

8.5 The bottom line on cost

Renewable energy systems are more expensive to purchase outright, which is one reason so many humanitarian actors continue to rely on diesel. Short-term funding cycles define most humanitarian crisis funding, creating a structural barrier to adopting renewable energy. However, although crisis funding is often short-term, many humanitarian crisis situations drag on for years. Given this grim reality, donors and humanitarian agencies need to take a longer-term view of their programming in South Sudan, making the case for renewable energy all the more compelling because the economic value of renewable energy is unlocked over time.

As was noted, the cost of diesel varies in different parts of South Sudan depending on access to foreign markets, the reliability of the supply chain, and the security situation, but diesel is expensive all over South Sudan relative to its neighbors and most of the world. In conflict-affected, landlocked parts of the country, such as Malakal, diesel is so expensive that a large-scale solar-plus-storage system to support humanitarian operations would pay for itself. The savings for a humanitarian agency on diesel fuel would pay for the full cost of the renewable energy system more quickly than in Juba, which has access to cheaper fuel by road via Kenya and Uganda. Given that solar systems have a lifespan of 20 years or more, the economic benefits of a solar system will only increase with time. In rural areas, it may take only 2–3 years to recoup the cost; in Juba, where diesel is less expensive, it may take 4–5 years.

At July 2017 prices, the cost of buying and installing a 650-kW solar-plus-storage system in South Sudan would be around US$1.8 million. The specific costs and economics of other projects across the country could be researched and assessed as part of an initial donor-supported assessment for solar transitions in the country.

8.5.1 How to pay for the transition

The dramatic growth of renewable energy around the world has been aided by a range of financing mechanisms, including tax credits and leasing schemes similar to home mortgages, that have allowed solar developers, businesses, and homeowners to pay down the cost of solar systems over time. Such credit or leasing facilities do not exist in South Sudan or in most other conflict-affected societies. Most humanitarian agencies and the vast majority of South Sudanese citizens do not have the resources to buy renewable energy systems outright.

Donors could help to fund a pivot toward renewable energy in several ways. One option would be to create a funding pool for outright purchase, rationalizing such a large, one-time up-front cost as the price for achieving much lower energy expenditures over time while creating a long-lasting energy infrastructure that also serves social service and peace-building goals. A second option would be to extend multiyear lease-like payment options to humanitarian grantees or to provide guarantees of multiyear funding so that individual organizations could seek their own financing. International solar developers were actively exploring South Sudan before the resumption of fighting, and a Norwegian company, Kube Energy, has successfully developed solar leasing business for humanitarian actors in South Sudan and the wider region. A third, complementary option is to create a financing mechanism specifically to support renewable energy in conflict and crisis settings, such as the Peace Renewable Energy Credit (P-REC).[4]

[4]The PREC was developed as a new variant of the Renewable Energy Credit (REC) mechanism. RECs represent 1 mWh of renewable energy generated and trade successfully in billion-dollar markets in North America and Europe, allowing both public and private sector actors to meet their renewable energy commitments by purchasing these virtual claims. PRECs could be generated from renewable energy

8.5.2 Maximizing benefits and mitigating risks

As has been noted, solar systems have been destroyed or looted in the civil war, a fact that may discourage donors and investors from funding a switch to renewable energy. The danger should be put in perspective, however; unlike in some other conflict settings, in South Sudan, solar panels are no more likely to be stolen than other means of power generation. Furthermore, some of the potential security risks can be mitigated thanks to the nature of the three settings discussed. The POC camps are located within UNMISS bases, behind fences and protected by UN peacekeepers. Although hospitals and other components of the country's health infrastructure have suffered, certain locations, such as the Juba Teaching Hospital, have survived, perhaps because they are recognized as safe and neutral spaces. The NGO compounds are primarily walled compounds with their own security measures, and solar panels could be installed on roofs to deter theft.

To enhance the sustainability of a solar initiative and to ensure that the South Sudanese benefit from the outset of a transition, new investment in renewable energy should be coupled with a significant commitment to fund local capacity building and training programs in solar energy. Donor support for such a transition would help to bring foreign solar developers to the country, create opportunities for the South Sudanese to get into the business of installing and maintaining solar systems, and provide a critical economic building block through electrification capacity. Increased competition would bring a variety of benefits, including lower prices.

8.6 Conclusion and recommendations

Renewable energy is not a solution to South Sudan's myriad challenges and will not resolve the problems that drive the conflict. However, current international humanitarian funding streams do provide an opportunity to chip away at these dynamics by building cleaner, cheaper, long-lasting energy infrastructure, whether in the form of small systems for individual NGO compounds, larger systems serving hospitals and health clinics, or extensive systems in POC camps.

projects in South Sudan, for example, sold back into the voluntary renewable energy markets to link existing renewable energy markets to fragile settings, and could create a new revenue stream to help promote clean energy in support of peacebuilding goals. For details, see the work of Energy Peace Partners at http://www.energypeacepartners.com and RAEL (2017).

International donors are understandably hesitant to invest in infrastructure projects in South Sudan, given that so much has been destroyed over the past few years, including significant international humanitarian assets. However, investing in renewable energy systems anchored in humanitarian activities differs from more traditional infrastructure investment in two key ways. First, donors are already funding expensive diesel systems for virtually all their humanitarian grantees. Transitioning to renewable energy would be a cost-saving strategy for future humanitarian programming. Second, these systems would generally be protected within contained compounds and could be mounted on containers or rooftops. Larger systems for POC camps would be located within the larger perimeter of the extended UN bases and thus would be well protected.

This solar energy infrastructure offers cleaner, cheaper, and long-lasting electricity generation and creates a new pro-peace asset. It would be a broader opportunity for South Sudan to escape its current development path, which depends exclusively on the petroleum sector. Rather than relying solely on the construction of megadams and a national grid to electrify the country, a scenario that, given the conflict and economic crisis, seems many years away even in the best-case scenario, solar power offers an easily scaled solution that works on and off the grid in both rural and urban settings.

An investment in renewable energy would provide tangible evidence to civilians and politicians alike of the country's opportunity to leap-frog older technologies and embrace a green development path that takes full advantage of technological developments and broader global political and investment interest in environmentally friendly strategies. Providing basic energy services through PREC investment, then, is a unique opportunity to jump-start peace building. Similar efforts include national government investments in trust funds such as the World Bank's Energy Sector Management Assistance Program and NGO campaigns that address landmines or gender equality.

The situation in South Sudan is one of the most challenging ones facing the international community today. New approaches are needed to help chart a path out of the crisis and to recapture the hope and optimism that defined South Sudan's future just a few years ago. International humanitarian donors can adapt to this longer-term lens, but doing so calls for new thinking and innovative approaches. Renewable energy presents a rare win-win opportunity, with benefits both in the short term and for years to come.

The fossil fuel– and renewable energy–rich nature of South Sudan means than many development paths are possible. The challenge is not assets but, as in many nations, management.

Current energy spending among government offices, as well as in UN and humanitarian agencies, must be assessed so that all participants create more transparency on energy costs and usage in South Sudan. Relief organizations must be encouraged to publish their actual energy supply and maintenance costs on a standard, levelized, cost-of-energy basis.

While more stability has come to South Sudan recently, this has been cited and hoped for many times before. A key step is to keep all parties—internal and international—in ongoing discussions concerning existing or new mechanisms to pool funds to help finance renewable energy systems for humanitarian grantees with multiple donors.

Efforts to implement pilot projects have seem some successes. These efforts, even a decade after independence, need ongoing to support and commitment to determine the relative economics and payback periods (length of time for diesel cost savings to equal cost of solar systems) for both small and large relief settings.

As in any resource-rich but governance- and commitment-poor state, there is a need to convene a global summit of humanitarian support groups to share the results of these pilot studies and encourage consideration of a transition to clean energy by the aid and relief community as a whole.

It remains key for internal parties to explore options and scenarios for transitioning the energy infrastructure for local benefit, to include maintaining donor involvement, including possibly ownership, in the protection and management of newly installed renewable energy systems as well as planning for new training and capacity-building programs to support South Sudan's solar sector.

Even with some degree of current stability, it is critical to commission independent groups to evaluate the planning issues that are involved in designing solar-plus-storage energy systems to become the new backbone of clean energy infrastructure as IDPs and refugees gradually move back to the towns.

References

Akec, J.A. (2016, December 11). South Sudan's economy: Is fuel the new dollar? *Sudan Tribune*. Retrieved from http://www.sudantribune.com/spip.php?article61080.

Alstone, P., Gershenson, D., & Kammen, D.M. (2015). Decentralised energy systems for clear electricity access. *Nature Climate Change*, *5*(4), 305–314.

Altai Consulting. (2014). South Sudan: Mapping the supply chain for solar lighting products. *Altai Consulting.* Retrieved from http://www.altaiconsulting.com/wp-content/uploads/2016/03/South-Sudan-Mapping-the-Supply-Chain_July-2014.pdf.

Bazilian, M., & Chattopadhyay, D. (2015). Considering power system planning in fragile and conflict states. *EPRG Working Paper 1518.* University of Cambridge.

Global Humanitarian Assistance. (2011). Resource flows to Sudan. *Global Humanitarian Assistance.* Retrieved from http://devinit.org/wp-content/uploads/2011/07/gha-Sudan-aid-factsheet-2011-South-Sudan-focus1.pdf.

GRSS. (2011). South Sudan Development Plan 2011–2013. *Government of the Republic of South Sudan.* Retrieved from http://www.grss-mof.org/wp-content/uploads/2013/08/RSS_SSDP.pdf.

Gueniat, M., Harjono, M., Missbach, A., & Viredaz, G.-V. (2016). *Dirty diesel: How Swiss traders flood Africa with toxic fuels.* Lausanne, Switzerland: Public Eye.

Hoth Mai, N.J., Mayai, A.T., & Tiltmamer, N. (2016). Sporadic fuel crisis in South Sudan: Causes, impacts, and solutions. *The Sudd Institute.* Retrieved from https://www.suddinstitute.org/assets/Publications/572b7eb2950f7_Sporadic-FuelCrisisInSouthSudanCausesImpacts_Full.pdf.

IEA. (2016). Key world energy statistics 2016. *International Energy Agency.* Retrieved from http://www.oecd-ilibrary.org/energy/key-world-energy-statistics-2016_key_energ_stat-2016-en.

Mansoor, M. (2016, October 6). Civil war spurs spike in solar energy use in Yemen. *Earth Island Journal.* Retrieved from http://www.earthisland.org/journal/index.php/elist/eListRead/civil_war_spurs_spike_in_solar_energy_use_in_yemen.

Nikolov, N. (2017, February 12). How sunshine is bringing radio to remote parts of South Sudan. *Internews.* Retrieved from https://www.internews.org/news/how-sunshine-bringing-radio-remote-parts-south-sudan.

Radio Tamazuj. (2017, July 30). Uganda to export electricity to South Sudan border towns. *Radio Tamazuj.* Retrieved from https://radiotamazuj.org/en/news/article/uganda-to-export-electricity-to-south-sudan-border-towns.

RAEL. (2017). RAEL holds first workshop on the Peace Renewable Energy Credit. *Renewable and Appropriate Energy Laboratory.* Retrieved from https://rael.berkeley.edu/2017/05/rael-holds-first-experts-workshop-on-the-peace-renewable-energy-credit.

REN21. (2016). Renewables 2016: Global status report. *Renewable Energy Policy Network for the 21st Century.* Retrieved from http://www.ren21.net/wp-content/uploads/2016/06/GSR_2016_Full_Report.pdf.

Stalon, J.-L., & Choudhary, B. (2017, June 27). Confronting climate change in South Sudan: Risks and opportunities. *Sudan Tribune.* Retrieved from http://www.sudantribune.com/spip.php?article62844.

Stevens, M. (2016, August 22). Kerry threatens to cut aid to South Sudan if peace isn't restored. *Wall Street Journal.* Retrieved from https://www.wsj.com/articles/kerry-threatens-to-cut-aid-to-south-sudan-if-peace-isnt-restored-1471881185.

Sudan Tribune. (2017b, December 3). S. Sudan quietly removes fuel subsidies over hard currency. *Sudan Tribune.* Retrieved from http://www.sudantribune.com/spip.php?article64157.

UN. (2014). Audit of fuel management in the United Nations Mission in South Sudan. *United Nations Report 2014/146.* Internal Audit Division.

UN. (2015). Budget performance of the United Nations Mission in South Sudan for the period from 1 July 2014 to 30 June 2015, Report of the Secretary General. *UN General Assembly Report A/70/599.*

USAID. (2017). South Sudan – Crisis fact sheet #5, Fiscal year 2017. *Relief Web.* Retrieved from https://reliefweb.int/report/south-sudan/south-sudan-crisis-fact-sheet-5-fiscal-year-fy-2017.

Wade, L. (2017). In Colombia, peace dividend for science proves elusive. *Science, 357*(6355), 958.

World Bank. (2016). HFS market surveys in South Sudan. *World Bank.* Retrieved from http://www.thepulseofsouthsudan.com/wp-content/uploads/sites/3/2016/08/MPS_v10.pdf.

Index

Note: Page numbers followed by "*f*" and "*t*" refer to figures and tables, respectively.

K

L

M